Advancing Sustainability through Green Chemistry and Engineering

About the Cover

The cover was designed by Rebecca L. Lankey and Paul T. Anastas, using a version of the logo of EPA's Green Chemistry Program as a motif.

ACS SYMPOSIUM SERIES **823**

Advancing Sustainability through Green Chemistry and Engineering

Rebecca L. Lankey, EDITOR
U.S. Environmental Protection Agency

Paul T. Anastas, EDITOR
White House Office of Science and Technology Policy

American Chemical Society, Washington, DC

Library of Congress Cataloging-in-Publication Data

Advancing sustainability through Green chemistry and engineering / Rebecca L. Lankey, editor, Paul T. Anastas, editor.

p. cm.—(ACS symposium series ; 823)

Includes bibliographical references and index.

ISBN 0–8412–3778–6

1. Environmental chemistry—Industrial applications—Congresses.

I. Lankey, Rebecca L., 1971- II. Anastas, Paul T., 1962- III. International Workshop on Green Chemistry and Engineering (4th : 2000 : Washington, D.C.) IV. Series.

TP155.2 .E58 2002
660′.028′6—dc21 2002016465

The paper used in this publication meets the minimum requirements of American National Standard for Information Sciences—Permanence of Paper for Printed Library Materials, ANSI Z39.48–1984.

Distributed by Oxford University Press

PRINTED IN THE UNITED STATES OF AMERICA

Foreword

The ACS Symposium Series was first published in 1974 to provide a mechanism for publishing symposia quickly in book form. The purpose of the series is to publish timely, comprehensive books developed from ACS sponsored symposia based on current scientific research. Occasionally, books are developed from symposia sponsored by other organizations when the topic is of keen interest to the chemistry audience.

Before agreeing to publish a book, the proposed table of contents is reviewed for appropriate and comprehensive coverage and for interest to the audience. Some papers may be excluded to better focus the book; others may be added to provide comprehensiveness. When appropriate, overview or introductory chapters are added. Drafts of chapters are peer-reviewed prior to final acceptance or rejection, and manuscripts are prepared in camera-ready format.

As a rule, only original research papers and original review papers are included in the volumes. Verbatim reproductions of previously published papers are not accepted.

ACS Books Department

Contents

Indexes

Preface

This volume is based on works presented at the Fourth Annual Green Chemistry and Engineering Conference, held at the National Academy of Sciences in Washington, D.C., June 27–29, 2000. The theme of the conference, "Sustainable Technologies: From Research to Industrial Implementation," was based on the premise that science and technology will need to play a major role if we are to move toward a society that is both environmentally and economically sustainable. It has been widely recognized that the field of green chemistry and engineering has resulted in remarkable achievements and notable positive impacts for human health and the environment during the course of the past decade. However, it may not be readily apparent to those not associated with the field that the positive impacts arising from these technologies are closely connected to meeting the broader goals of sustainability and sustainable development.

An important part of this greater understanding comes from knowing that green chemistry and engineering is based on addressing issues through pollution prevention and the minimization of hazards. The major challenges to sustainability, whether they be the proliferation of toxic substances, global climate change, or resource depletion, are all based on a particular basic hazard that can be addressed through the application of green chemistry and engineering. The speakers at the Green Chemistry and Engineering Conference and those selected for this volume were chosen on the basis of the relevance of their work to the issue of implementing sustainability.

The purpose of this book is twofold. The first is to demonstrate that the very specific work of fundamental science and engineering can have a broad positive impact on our global future. Often, this impact may be far beyond the imagination of even the inventors of a technology or the discoverers of a new molecule. This work will hopefully motivate scientists and engineers through the knowledge that their work can have an effect on sustainability. The second

Introduction

The current trajectory of society and the world is unsustainable. If one looks at any of the five major challenges to sustainability, which are population growth, energy production, food production, global climate change, and resource depletion, it is clear that the current practices of society cannot be maintained indefinitely.

While it took all of human history up until 1927 to reach a population of two billion people, it took only 72 years to triple that number and add another four billion people (*1*). Another billion will be on the planet in the next decade, with China contributing as many as the current population of the United States in that time. With a population that is expected to reach 9 billion around the year 2050 (*1*), the planet is showing signs of strain.

The empirical data of population growth show a dramatic correlation between a higher standard of living and sustainable population growth. Of the 78 million people that are added to the world each year, 95% live in the less developed regions (*1*). This fact highlights one of the core challenges for sustainable development:

- Higher standards of living are linked to stable population growth, but
- Higher standards of living historically have been accompanied by negative impacts on human health and the environment; so
- We must discover how to increase quality of life, especially in developing nations, while also minimizing the impact on the planet.

This challenge then comes largely to the scientists and engineers, and Green Chemistry and Engineering technologies will play a significant role in their solutions.

Supplying enough food for a population of six billion is less of a challenge than the distribution of food supplies. However, current agricultural practices are unsustainable (*2*), and the loading capacity of the planet will be challenged as population increases this century. Just as the depletion of fossil fuels is inherently unsustainable, the attempts at mitigating the environmental effects of their use will become economically unsustainable in the long term. The issues of global resource depletion and the diffusion of persistent and toxic chemicals have reached a point where there is an acceptance that the entire population of the earth cannot live in the same manner as the industrialized nations. For example, in 1996 the "ecological footprint" (a quantitative estimate of human pressure on global ecosystems) of OECD member countries was over three times that of non-OECD countries (*3*). It has been estimated that for 5.8 billion people to live at current North American standards would require the resources of three earth equivalents; a population of 10 to 11 billion could require six earth equivalents (*4*). Alternatively, it has been estimated that over the long term the earth can support only two billion people at a middle-class standard of living (*5*).

It should be remembered, yet not overstated, that the world has previously engaged in unsustainable practices that it successfully managed. When horses were the main source of transportation, it was predicted that cities would be literally buried beneath accumulated manure if trends continued. With the advent of new modes of transportation, the predicted fate of mankind was averted. Now, however, the challenges are even more fundamental. There are those that feel that efficiency and productivity are essential elements to a sustainable future. With the industrial revolution there was a tremendous increase in productivity on a per person basis. However, overall resource use was greatly increased in absolute terms (*2*). The recent information technology revolution that is still ongoing has brought about tremendous productivity gains and efficiencies related to the use of materials. However, these gains are marginal relative to the orders of magnitude improvement that is needed.

What can molecular science and engineering contribute to addressing the challenges to sustainability? The science and technology presented in this book provide some case studies that can argue that the necessary innovations to achieve sustainability will come at the molecular level. The current threats to the health of the biosphere and the systems that support it are either derived from or can be greatly mitigated for the most part by intrinsic properties within chemical structure.

The impact of designing materials and molecular transformations such that they are more environmentally benign extends throughout the lifecycle from materials input to end-of-life and beyond. The "beyond" incorporates the ability to design molecular structure to allow for the reincarnation, reuse, or deliberate degradation into innocuous products.

The empirical data (see Table I) suggest that an effective way to decrease population growth to sustainable levels is by increasing the quality of life of those in the developing world. However, it is essential that this be done in a manner that drastically reduces the amount of materials needed for that increase in quality of life. This challenge falls directly on the shoulders of chemists, chemical engineers and those involved in the enterprise of molecular design. It is only through innovations in materials and in energy generation and storage, for example that the type of efficiency and safety required on the order-of-magnitude level can be achieved.

There are few things that chemists and chemical engineers like more than a good challenge. In many ways, this is a wonderful time in history to be a chemist or chemical engineer. The challenges that confront the planet are nothing short of the most consequential that humanity has ever faced. Simply put, if we do not meet the challenges of maintaining the viability of the biosphere, humanity will not survive to see any future challenges.

Table I. United Nations Population Growth Estimates

Major Area	*1999 Population, (millions)*	*Population growth rate, 1995–2000 (%)*
World	5978	1.3
More developed regions	1185	0.3
Less developed regions	4793	1.6

SOURCE: Population Division of the Department of Economic and Social Affairs of the United Nations Secretariat, *The World at Six Billion*. (ESA/P/WP.154), 12 October 1999.

Challenges to Sustainability

Efforts in Green Chemistry and Engineering need to be centered on the major issues for sustainability. These include:

- Global climate change,
- Energy production,
- Food production,
- Resource depletion, and
- Toxics in the environment.

The principles of Green Chemistry (*6*) can be particularly useful in addressing these major challenges through the design of next-generation products and processes.

Global Climate Change

Global climate change is now recognized as a major issue of increasing concern. Energy generation, storage, and transport, discussed below, will continue to be a major factor in addressing the problem. Alternative, non-combustion sources of heat and power will need to be developed. However, the utilization of carbon dioxide also needs to be included in the toolbox of solutions to climate change. Chemists and chemical engineers have demonstrated the ability to use carbon dioxide in materials ranging from polymers to cement in value-added ways that enhance performance. While these applications capture significant amounts of carbon dioxide on a product- or process-specific basis, it is still a small part of the global problem and must only be viewed as a technological beginning. New carbon dioxide utilization technologies that use high-pressure carbon dioxide to help extract natural gas more efficiently not only provide a value-added service, but also immobilize large quantities of carbon dioxide that react deep in the wells to form carbonates. In other applications, road surfaces and other parts of our infrastructure could be made to react with carbon dioxide to form carbonates in ways that increase performance and even "heal" cracks. The important principle is to understand that as long as carbon dioxide is the essential byproduct of combustion processes, a use for this substance must be found so it is no longer a "waste" but is instead a useful feedstock.

Energy Production

For the immediate term, it is important that current technology strive to make the dominant fossil fuel sources as clean as they can be. However, it needs to be recognized that our long-term energy sources must be renewable to be sustainable, as a matter of definition. It will be the chemists and engineers that will make the photovoltaic materials needed for the economic feasibility of solar power, for example. It will also be the chemists and engineers who turn the promise of hydrogen fuels cells into a manufacturing reality. The issues of energy generation, storage, and transport are all intrinsically linked to the materials that are used, and these materials need to be designed and manufactured using the principles of Green Chemistry and Green Engineering.

Food Production

The manner in which food is produced can be as important in the long term as the quantity. While world food production is sufficient to feed the current population, distribution inefficiencies result in continued regional episodes of starvation. The current practices of agriculture are not sustainable. Green Chemistry and Engineering advances have resulted in the development of pesticides that are very target-specific and do not persist in the environment. In addition, the development of new fertilizers and fertilizer adjuvants has greatly decreased the amount of fertilizer needed, thereby reducing run-off that contaminates both surface and ground water. Green Chemistry and Engineering advances in agricultural chemicals also have the potential of allowing for more efficient production in regions of the world in the greatest need.

Resource Depletion

The human population is currently consuming the Earth's resources at an unsustainable rate. Some of the most advanced materials and life-saving medicines are derived from petroleum-based feedstocks, and yet we continue to burn petroleum for fuel. The depletion and dissipation of rare minerals also cannot be sustained by current practices. Green Chemistry principles are resulting in the development of material-efficient and renewable products and processes. Simply optimizing existing technologies may result in improvements of 10 to 25%, or in some cases even 50%, but generally this will not achieve the types of four-fold and ten-fold improvements necessary for sustainability. Improvements of this magnitude must come from a redesign at the molecular level and through the utilization of renewable materials including agricultural and other biowastes, carbon dioxide, and materials such as chitin. Nanoscience applied to nanotechnology will need to supply the functions needed by society with dramatic decreases in materials use.

Toxics in the Environment

The principles of Green Chemistry outline the design of chemical products and processes for reducing the intrinsic hazard of materials. Through appropriate molecular design, characteristics such as toxicity, persistence in the environment, and bioaccumulation can be minimized and manipulated the same as any other physical property of chemicals and other materials. Just as one can design a dye to be red, it can be designed so that it does not cause cancer. In the same way a plastic can be designed to be flexible, it can be designed to ensure that it does not cause endocrine-disrupting effects. This type of design requires the same kind of knowledge, skills, and techniques that chemists have been using for generations. However, it also requires a new perspective: the perspective that minimized intrinsic hazard, whether physical, global, or toxicological, is simply another performance criterion that must be incorporated.

Approaches

The individual chapters in this book demonstrate some creative thinking in turning the goals of sustainability into reality through the use of elegant science and engineering. An exciting aspect of green chemistry and engineering is that the field encompasses many types of materials, products, and processes (see, for example, *7, 8*). Green Chemistry and Engineering technologies provide both incremental and entirely innovative solutions to technological questions and demands, and these technologies also have demonstrable environmental and economic benefits.

One area of active investigation is that of catalysis, most notably addressed in this book by Collins and coauthors (Chapter 4), by Theopold and coauthors (Chapter 6), by Weinstock and coauthors (Chapter 7), and by Gaynor and coauthors (Chapter 9). Because of the well-recognized economic advantages of catalysis, the benefits offered to environmental sustainability may have been overlooked or understated. When a truly catalytic reaction replaces a stoichiometric reaction, potentially immense material savings can be realized. When a catalyst allows a reaction to proceed in a more selective manner, it results in less waste byproduct and therefore greater resource efficiency. Just as importantly, the lowering of the energy necessary in a chemical transformation can have dramatic benefits, especially for large-scale processes.

The important areas of biofeedstocks may well be one of the most important areas of Green Chemistry and Engineering research. This area affects topics ranging from agriculture to global climate change to energy production. Gao and coauthors (Chapter 11) describe the production of biocatalysts and the conversion of biomass to chemicals. The chapters by Wyman (Chapter 3) and by Wool and coauthors (Chapters 13 and 14) illustrate the scientific and technological advances surrounding biofeedstocks.

Applications and Case Studies

The Green Chemistry and Engineering technologies described in this book illustrate specific ways that the ideals and goals of sustainable development and industrial ecology can be implemented. In Chapter 2, Fiksel explores how the concept of industrial ecology provides a useful systems perspective to support sustainable development while also assuring shareholder value creation. He describes how the holistic framework of industrial ecology can guide the transformation of industrial systems from being a linear model to being a closed-loop model that more closely resembles the cyclical flows of ecosystems. More specifically, Fiksel explains how the field of Green Chemistry and Engineering represents an important set of design strategies for achieving the goals of industrial ecology through fundamental innovations.

Wyman (Chapter 3) addresses the needs for future energy production and sustainable resource use by examining lignocellulosic biomass as a low-cost and sustainable resource that can be used for the large-scale production of organic fuels and chemicals. The conversion of biomass for these purposes can help to decrease the amount of carbon dioxide and other gases thought to contribute to global warming while simultaneously decreasing the use of petroleum-based chemicals and products. Avdances in biomass conversion technologies can also provide important economic and strategic advantages.

Reducing the dissipation of toxic and hazardous chemicals in the environment through the use of catalysis is the focus of Collins and coauthors (Chapter 4). Their work describes the catalytic activation of hydrogen peroxide, and this technology promises valuable benefits for the pulp and paper industry.

Their catalysts, designed to activate hydrogen peroxide to rapidly bleach wood pulp at high selectivity at relatively low temperatures, provide a new totally chlorine-free approach. Furthermore, the catalyst-activated peroxide can be used to remove "color" in chlorine dioxide bleaching plant effluents using low peroxide and minute catalyst concentrations. This chapter summarizes potential applications of this technology that promise to improve capital and operating costs, product quality, environmental performance, and energy efficiency.

The environmental issues surrounding agriculture include the run-off of chemicals into groundwater, and Thompson and Sparks (Chapter 5) detail one green chemistry solution to help alleviate this problem. They describe how a unique chemistry produced by fermentation of a naturally occurring soil microorganism is providing a greener solution for agriculture, an integrated pest management (IPM) technology that incorporates high levels of efficacy but with benign environmental effects. The IPM technology is an effective tool for producing more food and fiber on fewer acres while also protecting the environment, because the compound does not leach, bioaccumulate, volatilize, or persist in the environment. The described technology demonstrates how food production can be made more environmentally friendly.

Theopold and coauthors (Chapter 6) use catalysis to address the need in industry to decrease the generation of harmful byproducts and to utilize waste streams, especially those that contain gases thought to contribute to global warming. An attractive environmental alternative is to use oxygen and nitrous oxide as terminal oxidants. However, a problem with this approach is the kinetic inertness of these molecules. Theopold and coauthors are working to solve this problem by activating dioxygen and nitrous oxide by binding to transition metals and ultimately creating homogeneous catalysts to perform selective oxidations.

The research of Weinstock and coauthors (Chapter 7) also uses catalysis, with a goal of decreasing the use and dissipation of toxic and hazardous chemicals in the environment. Their environmentally benign technology is based on a solution of equilibrated polyoxometalate salts and oxygen, and a primary application is the bleaching of wood pulp in the manufacture of paper. Their research describes a new and general approach to designing soluble transition metal catalysts for aerobic oxidation in water. Unlike conventional technologies, this chemistry uses only oxygen and water and produces no liquid waste, thus eliminating the many health and environmental problems that have been associated with the use of chlorine compounds.

The efficiency of food production can be increased with fewer environmental stressors, as shown by Georgis and coauthors (Chapter 8). The studied compound, thermal polyaspartate, is commercially available as a nutrient absorption enhancer in agricultural applications. Their investigations have shown that certain variations of thermal polyaspartate can enhance the uptake and effectiveness of insecticides and herbicides. The practical benefits of this technology have been demonstrated successfully in greenhouse and field tests.

Gaynor and coauthors (Chapter 9) have developed the technology of atom transfer radical polymerization, a system that allows for the synthesis of well-defined polymers with predetermined molecular weights, functionalities, and architectures. Being able to control material characteristics makes it possible to prepare polymers with novel uses. Materials made with this technology perform more efficiently than comparable polymers, require less material to manufacture, and generate less catalytic waste than conventional processes. Furthermore, these polymerizations can be carried out in environmentally friendly media such as water, supercritical or liquid carbon dioxide, and ionic liquids. This technology contributes to the conservation of resources and also to the use of chemical processes that are less hazardous than conventional methods.

One of the first steps towards implementing a new technology is designing it. As mentioned above, designing a product or process with the goal of pollution prevention in mind is a fundamental principle of Green Chemistry. To aid in this design process, Hendrickson (Chapter 10) presents a rigorous system of synthesis design that helps to discover all the best syntheses for any given target compound. Hendrickson's SynGen program executes the design model, based on skeletal dissection and characterization of compounds and reactions. The use of such design programs is especially useful to help avoid the unnecessary use of resources, especially when the use of potentially hazardous chemicals cannot be avoided.

Gao and coauthors (Chapter 11) explore the use of catalysis for the conversion of biomass to industrial chemicals and the production of industrial biocatalysts. Using a genetically modified starch-degrading *Saccharomyces* strain, the expression of a bacterial enzyme is investigated in starch media, which include starch from corn and from potato waste. The ability to use biomass as a feedstock is valuable both from the standpoint of utilizing the waste and of decreasing the use of the toxic and hazardous chemicals used in conventional production processes.

To decrease the use of hazardous chemicals, Berejka (Chapter 12) has studied the use of ionizing radiation generated from electron-beam accelerators and the use of photoinitiated reactions started by high intensity UV light. These technologies have helped industry to eliminate the need for volatile organic compounds from coatings, inks, and adhesives. These processes also are energy-efficient and minimize wastes, which conserves both energy and materials. Ionizing radiation can also be used to decrease the amount of waste sent to landfills by precipitating sulfur dioxide and nitrous oxides from stack gases and by dehydrohalogenating toxic halocarbons found in wastewater and in soils.

The use of biobased materials is being implemented by Wool and coauthors (Chapters 13 and 14). They have studied the use of new and improved materials made from renewable resources, and these materials can biodegrade or be recycled. High-performance, low-cost materials can be made using plant oils, natural fibers, and lignin. When these bio-based materials are combined with natural fibers, glass fibers, and lignin, new low-cost composites are produced that are economical in many high-volume applications. These composites are used in agricultural equipment, automotive sheet molding compounds, civil and rail infrastructures, marine applications, and the construction industry. This research not only supports decreasing the use of non-renewable resources and energy consumption, but also can help to decrease levels of carbon dioxide when additional biomass is grown to be used in these applications.

As discussed above, fossil fuels are not a viable long-term solution for a sustainable world. However, in the interim until sustainable energy technologies are fully developed, green chemistry and engineering can help to make the technologies currently in use more environmentally friendly. The research of Maroto-Valer and coauthors (Chapter 15) focuses on the development of a technology to simultaneously reduce emissions and convert byproduct waste streams into value-added products. For example, the described technology combines the installation of low-nitrogen-oxide burners that efficiently reduce nitrogen oxide emissions, along with strategies that help to manage the associated increase in byproduct streams, which are mainly fly ash and unburned carbon. These technologies are addressing today's needs in the U.S. electric power industry by helping to decrease the emission of nitrogen oxides from coal combustion furnaces and their associated byproduct waste streams.

Conclusions

The ability to invent a sustainable future exists in the talents of our chemists and engineers. The two elements that are still needed are education and a sense of urgency. We need to expand our understanding of the molecular basis of hazard into the current chemistry curriculum. The seemingly miraculous ability of our chemists and chemical engineers to "engineer away environmental problems" after they are formed must now be reoriented to include the design and implementation of processes that obviate the need for the utilization of hazards wherever feasible. This education, this action, must not be undertaken at a pedestrian pace. It needs to be done now, with the sense of urgency that the sustainability of our planet warrants.

There are those who argue that sustainability is far more than a scientific and technological challenge, that it involves complex social interactions and value systems. They may be correct. However, while they endeavor to change the hearts, minds, and behavior of six billion people to make society more sustainable, it is the responsibility of the small percentage of the population that is knowledgeable in science and engineering to make current (and likely future) behavior as sustainable as possible. This can be done through the design of intrinsically benign products and processes. It is not a feasible option simply to tell the developing world that increasing their quality of life is an unrealistic expectation. To make that increased quality of life have a minimal negative impact on the Earth is our greatest challenge, and an attainable one.

References

1. Population Division of the Department of Economic and Social Affairs of the United Nations Secretariat, *The World at Six Billion;* (ESA/P/WP.154); United Nations: New York, October 12, 1999.
2. OECD Environment Directorate, *OECD Environmental Outlook,* Organisation for Economic Co-operation and Development: Paris, 2001.
3. *Living Planet Report 2000.* Loh, J., Ed. WWF–World Wide Fund For Nature: Gland, Switzerland, October, 2000.
4. Rees, W.; Wackernagel, M. Ecological Footprints and Appropriated Carrying Capacity: Measuring the Natural Capital Requirements of the Human Economy. In *Investing in Natural Capital: The Ecological Economics Approach to Sustainability;* Jansson, A.-M., Hammer, M., Folke, C., Costanza R., Eds.; Washington, DC: Island Press, 1994; pp 362–390.
5. Mckibben, B. A Special Moment in History. *The Atlantic Monthly,* May **1998,** *281* (5), pp 55–78.
6. Anastas, P. T.; Warner, J. C. *Green Chemistry: Theory and Practice;* Oxford University Press: New York, 1998.
7. *Green Chemical Syntheses and Processes;* Anastas, P. T., Heine, L. G., Williamson, T. C., Eds; ACS Symposium Series 767; American Chemical Society: Washington, DC, 2000.
8. *Green Engineering;* Anastas, P. T., Heine, L. G., Williamson, T. C., Eds.; ACS Symposium Series 766; American Chemical Society: Washington, DC, 2001.

Chapter 2

Sustainable Development through Industrial Ecology

Joseph Fiksel

Battelle Memorial Institute, Life Cycle Management Group, 505 King Avenue, Columbus, OH 43201

Sustainable development has emerged as a powerful force that is reshaping the policies and practices of global manufacturing firms. The principle of sustainability states that as we strive for economic prosperity and growth, we must not compromise the quality of life for our descendants. For corporations, seeking sustainability involves designing environmentally and socially responsible technologies, products, and processes with a full awareness of their life cycle costs and benefits. The concept of industrial ecology (IE) provides a useful systems perspective to support sustainable development while assuring shareholder value creation. IE is a broad, holistic framework for guiding the transformation of industrial systems from a linear model to a closed-loop model that resembles the cyclical flows of ecosystems. In particular, the field of green chemistry represents an important set of design strategies for achieving the goals of IE through fundamental innovations.

Emergence of the Sustainable Development Concept

The original concept of "sustainable development" (SD) set forth a vision of industrial progress that respects both human needs and global ecosystems, assuring the quality of life for future generations. The underlying principle was first introduced by the Bruntland Commission in 1987, which defined SD as

> ...industrial progress that meets the needs of the present without compromising the ability of future generations to meet their own needs.
> —World Commission on Environment & Development (*1*)

There are a number of pervasive issues associated with SD that present severe challenges to the continued growth of global industries, particularly in developing nations. According to the World Bank (*2*), these include:

- Potential climate change due to CO_2 and other global warming gases;
- Degradation of air, water, and land in industrialized areas;
- Depletion of natural resources, including fresh water, biomass, and minerals;
- Loss of agricultural land due to deforestation and soil erosion;
- Threatened wildlife habitats, including forests, reefs, and wetlands;
- Lack of potable water for approximately 1.5 billion people;
- Unsanitary urban conditions (over 2 billion people lack access to sewers);
- Proliferation of both viral and bacterial infectious diseases;
- Increasing resource needs due to population growth (~90 million/yr);
- Social disintegration resulting from displacement of traditional lifestyles;
- Growing income gaps between rich and poor strata of society;
- Lack of primary education for about 130 million children worldwide; and
- Extreme poverty (about 3 billion people, or roughly half the world's population, are estimated to earn under U.S.$2/day).

For the most part, these problems are neither speculative nor localized. Collectively, they represent the greatest threat that human society has ever faced, easily dwarfing public health concerns over environmental carcinogens in industrialized parts of the world. Historically, most of these issues appeared remote from the comfortable perspective of the more affluent nations. Today, thanks to the revolution in modern telecommunications, the globalization of markets, and the rapid growth of developing economies, SD issues cannot be so easily ignored.

A landmark event that raised the international visibility of SD was the Earth Summit of 1992, held in Rio de Janeiro, officially known as the United Nations Conference on Environment and Development. The issues discussed there transcended national and industrial boundaries, ranging from the export of pollution to developing nations to the international equity of environmental regulations and the sustainability of population and industrial growth in the face of limited planetary resources (*3*). In addition to the SD principle above, a number of key principles were articulated at the Rio Summit, including:

- The "precautionary approach" that is, where there are threats of serious or irreversible damage, scientific uncertainty shall not be used to postpone cost-effective measures to prevent environmental degradation.
- The notion that environmental protection should constitute an integral part of the development process and cannot be considered in isolation from it.
- The "polluter pays" concept, that is, polluters should bear the cost of pollution.

There is now increasing support around the world for public policy measures to promote SD, including not only regulatory requirements, but also establishment of waste reduction targets, promotion of "cleaner" production methods, and government support for research and development into environmentally sound technologies to stimulate the practice of SD.

The Response of Industry

The threat of government decision-makers assigning pollution taxes to selected industrial sectors was certainly a forceful "wake-up call" to the business community. For example, the United Kingdom recently enacted an energy tax (known as the Climate Change Levy) that penalizes industrial firms if they do not achieve certain levels of energy efficiency. However, apart from any government mandates or incentives, there has been a quiet revolution in industry attitudes toward the SD issues raised in Rio. A significant factor in this revolution was the creation of the World Business Council for Sustainable Development (WBCSD), a Geneva-based consortium of over 100 leading companies formed in 1990 to develop a global perspective on SD. Their book, *Changing Course*, is an important manifesto describing both the challenges and the opportunities for profitability associated with corporate sustainability (*4*). Subsequently, WBCSD has published a series of studies that demonstrate the business value of sustainability and is coordinating the definition of agendas for change in industries such as pulp and paper, mining, cement, transportation, and electric power.

SD is arguably a logical outgrowth of evolving industry practices, including environmental stewardship, pollution prevention, waste minimization, and design for environment methods, which seek to replace "end-of-pipe" pollution control with more cost-effective process improvements. However, there are two important differences that distinguish SD from historical environmental management practices:

- Introduction of the socio-economic dimension, focusing upon the beneficial or adverse impacts of industrial growth upon societal well-being.
- Recognition of the linkages between SD and competitive advantage in the marketplace, which elevates these issues to a strategic level.

Confronted by rising stakeholder concerns, the business community has recognized that long-term success depends not only on financial performance, but also on social and environmental performance. Today, we define a "sustainable business" as one that is able to anticipate and meet the needs of present *and future* generations of customers and stakeholders, encompassing three dimensions of need known as the "triple bottom line":

- Economic prosperity and continuity for the business and its stakeholders.
- Social well-being and equity for both employees and affected communities.
- Environmental protection and resource conservation, both local and global.

From this perspective, corporate sustainability is not just altruism—rather, it is an enlightened response to emerging market forces. Instead of merely listening to the voice of the individual customer, companies are beginning to listen to the collective voice of the larger Customer; namely, human society. Public awareness of sustainability creates opportunities for companies to address changing stakeholder expectations with new technologies, products, and services, thus gaining competitive advantage (*5*).

Sustainable Business Practices

In recent years, a number of leading multinational corporations have established highly visible sustainability programs. For example, top management at BP Amoco, Dow, DuPont, Ford, General Motors, Royal Dutch Shell, and IBM have gone public with ambitious commitments to generate shareholder returns while addressing the needs of humanity. In general, the common purpose of these programs is to shift the company operations from a traditional, resource-intensive, and profit-maximizing business model to a more eco-efficient, socially responsible, and value-maximizing model. This shift aligns nicely with the financial goal of increasing shareholder value by raising profits while reducing the cost of capital—that is, doing more with less.

Early adopters of SD have included companies in many different industries, such as chemicals, consumer products, pharmaceuticals, motor vehicles, computers and electronics, forest products, petroleum, and even floor-coverings. Rather than following a prescribed approach, each company has explored how it can integrate sustainability into its own business strategy. For example, DuPont has adopted a company-wide sustainability indicator (shareholder value added per pound of product) that reflects their overall goal of creating greater value with fewer resources—again, doing more with less. In contrast, Procter & Gamble (P&G) is building its sustainability efforts around two focused themes:

- Water—because 85% of its products involve household water use, P&G is investigating how to enhance both water conservation and water quality.
- Health & hygiene—P&G is seeking to enhance the global contributions of its products to sanitation, health care, infant care, and education.

Typically, there are several levels of sustainable business practices, with increasing levels of difficulty. The most basic level is corporate initiatives such as philanthropic programs aimed at solving sustainability problems. The next level often involves reducing the "ecological footprint" associated with the product life cycle, including manufacturing, use, and end-of-life disposition. The most challenging level is enhancing the *inherent social value* created by the firm's operations, products, and services, which may range from assuring human health and nutrition to stimulating consumer education and growth of new businesses. At this level, challenging trade-offs may arise—for example, balancing job creation and economic development against community concerns about industrial pollution and environmental justice.

One of the greatest pitfalls of SD is viewing it as a regulatory compliance issue and addressing it through typical environmental, health, and safety programs. Companies that are successful SD practitioners typically view SD as a strategic issue, have the Chief Executive Officer or other senior executives as champions, and encourage cross-functional collaboration throughout the organization in pursuit of innovative products and business models. For example, product designers compare the sustainability profiles of competing product concepts, marketers analyze how to differentiate their product or service in terms of life-cycle cost of ownership, production managers apply environmental accounting to quantify hidden environmental costs, and strategic planners assess the consequences of long-term SD scenarios. Finally, most practitioners agree that for SD to become truly integrated into business decisions, a systematic SD performance measurement process is essential.

Four fundamental principles can help companies address the challenges associated with measuring and reporting product sustainability (*6*). These are:

- Address the dual perspectives of *resource* consumption and *value* creation.
- Include economic, environmental, and societal aspects.
- Systematically consider impacts associated with each stage in the product life cycle, including resource extraction, procurement, transportation, manufacturing, product use, service, and disposition.
- Develop both *leading* indicators to measure internal process improvements and *lagging* indicators to measure external results.

Most companies that are committed to SD have established measurable performance goals and have expanded their annual environmental, health, and safety progress reports into sustainability reports aimed at stakeholder communication. Tools such as life-cycle assessment are increasingly used to measure performance in terms of total societal impacts. Meanwhile, emerging consensus standards such as the Global Reporting Initiative (*7*) are helping to establish key performance indicators that will support industry benchmarking. Examples of commonly used SD indicators are shown in Table I.

Table I. Examples of Sustainability Indicators

Economic	*Environmental*	*Societal*
Direct • Raw material costs • Labor costs • Capital costs • Operating costs	**Material Consumption** • Product & packaging mass • Useful product lifetime • Hazardous materials used	**Quality of Life** • Breadth of product availability • Knowledge enhancement • Employee satisfaction
Potentially Hidden • Recycling revenue • Product disposition cost	**Energy Consumption** • Life cycle energy • Power use in operation	**Peace of Mind** • Perceived risk • Complaints
Contingent • Employee injury cost • Customer warranty cost	**Local Impacts** • Product recyclability • Impact upon streams	**Illness & Disease Reduction** • Illnesses avoided • Mortality reduction
Relationship • Loss of goodwill due to customer concerns • Business interruption due to stakeholder interventions	**Regional Impacts** • Smog creation • Acid rain precursors • Biodiversity reduction	**Accident & Injury Reduction** • Lost-time injuries • Reportable releases • Number of incidents
Externalities • Ecosystem productivity loss • Resource depletion	**Global Impacts** • CO_2 emissions • Ozone depletion	**Health & Wellness** • Nutritional value provided • Food costs

Industrial Ecology as a Normative Framework

One of the most significant barriers to sustainable development is the broad scope and complexity of the issues that need to be addressed, ranging from minimization of ecological impacts associated with industrial waste streams to enhancement of quality of life in developing nations. It is difficult for decision-making teams to anticipate the multitude of cause and effect chains that drive the ultimate beneficial or adverse impacts of their proposed investments or technical innovations. Instead, they need to adopt fundamental *design principles* that are "normative" in the sense that they will generally lead to positive outcomes, although one can never guarantee the absence of hidden drawbacks.

As shown in Figure 1, during the late 20th century there was a rapid evolution in normative approaches used by industry managers for integrating environmental and social considerations into business decision-making.

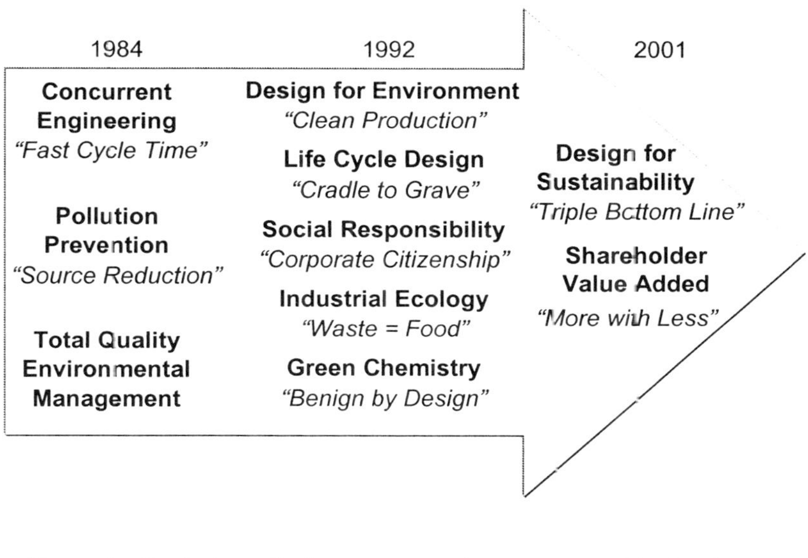

Figure 1. Evolution of industry approaches to sustainable development.

The convergence of accelerated product development methods (such as concurrent engineering) with pollution prevention thinking led to the widespread adoption of approaches such as Design for Environment, which provides systematic guidelines for anticipating and enhancing the environmental performance of new products and processes. This was accompanied by the development of analytical approaches for evaluating performance at both a product and a system level, most notably the body of methods known as *life-cycle assessment.* An early foundation for industrial ecology was Robert Ayres' work on industrial metabolism, which examined the life-cycle patterns of energy and materials in an industrial system (*8*). These interrelated threads of thought have evolved into what are now the two dominant themes of sustainable business—designing sustainable systems and creating shareholder value.

In the above context, industrial ecology is a relatively new discipline that offers design principles to support both business innovation and evaluation of sustainability performance. Stated simply, industrial ecology is *a broad, holistic framework for guiding the transformation of industrial systems from a linear model to a closed-loop model* that resembles the cyclical flows of ecosystems. In nature, there is no waste—one creature's waste becomes another creature's food. Thus, IE provides a foundation for rethinking conventional product or process technologies and discovering innovative pathways for reuse and recovery of industrial waste streams. However, it should be noted that industrial ecology focuses mainly on the ecological and economic dimensions of sustainability. A comparable paradigm that supports societal sustainability considerations has not yet been developed.

There are as many alternative definitions of industrial ecology as there are authors promoting the concept. Here are some examples:

> Industrial ecology is the study of the flows of materials and energy in industrial and consumer activities, of the effects of these flows on the environment, and of the influences of economic, political, regulatory, and social factors on the flow, use, and transformation of resources.
>
> —Robert M. White, Past President, U.S. National Academy of Engineering (*9*)

> Industrial ecology . . . is a systems view of industrial operations in which one seeks to optimize the total materials cycle from virgin material, to finished material, to component, to product, to waste product, and to ultimate disposal. Factors to be optimized include resources, energy, and capital.
>
> —David Chiddick, Vice President, AT&T (*10*)

> Industrial ecology takes the pattern of the natural environment as a model for solving environmental problems, creating a new paradigm for the industrial system in the process.
>
> —Hardin Tibbs (*11*)

> [T]he ultimate goal . . . is bringing the industrial system as close as possible to being a closed-loop system, with near complete recycling of all materials.
>
> —Ernest Lowe (12)

All of these definitions share some basic assumptions, namely that industrial systems must function within natural constraints, that ecosystem principles offer guidance in the design and management of industrial systems, and that optimization of material and energy flows will generate both environmental and economic benefits

One influential group that has stimulated the introduction of IE thinking into major corporations is the Natural Step (*13*), a Swedish-based organization that advocates four required "system conditions" for achieving sustainability:

- Avoid increases in substances extracted from the Earth's crust.
- Avoid increases in non-degradable substances produced by society.
- Avoid displacement, over-harvesting, or other ecosystem manipulation.
- Use resources fairly and efficiently to meet human needs globally (this is considered an imperative for achieving the first three conditions).

While these conditions are challenging to meet in the absolute sense, they are consistent with the general principles of IE articulated above.

Cyclical Flows in Industrial Systems

At the heart of industrial ecology is the notion of emulating natural systems by seeking cyclical industrial flows, which some have called "biomimicry" (*14*). The traditional, linear model of industrial economics assumes that raw materials are acquired at the beginning of the value chain and are then processed and incorporated into intermediate and final products, which are in turn delivered to product users or consumers, who use the products and discard or recycle them. Wastes that are generated along this chain are "exported" outside the system for appropriate disposition. The notion of *eco-efficiency*, in its simplest form, focuses on the manufacturing stage of the value chain and seeks to generate higher-value outputs with reduced resource inputs (see Figure 2).

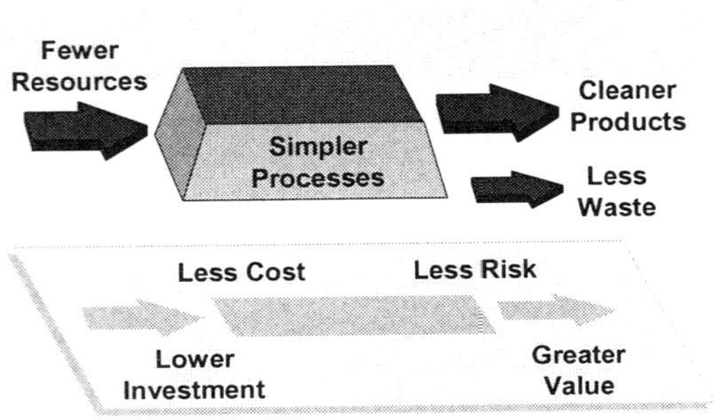

Figure 2. Benefits of increasing eco-efficiency for a manufacturing activity.

In contrast, the cyclical model approaches industrial activities from a systems perspective and seeks to optimize overall efficiency, ultimately eliminating all waste. This concept is illustrated in Figure 3, which shows several different types of pathways for waste utilization, including:

- Closed-loop recovery (recycling of feedstock process waste).
- Conversion of process waste to byproduct (from Product A to Product B).
- Recycling of post-consumer waste (from Product A consumption).
- Waste-to-energy conversion (Product A post-industrial and post-consumer).

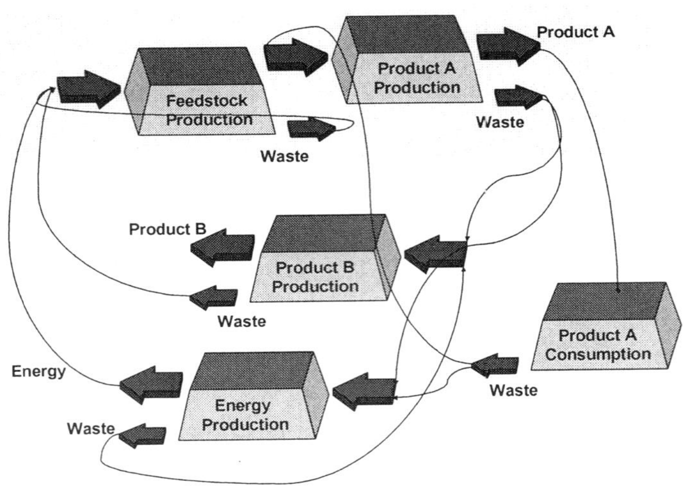

Figure 3. Examples of cyclical flows in an industrial ecosystem.

Each of the industrial ecology flows depicted in Figure 3 can potentially provide economic benefits to both the waste generator and the waste consumer.

An example of how industrial ecology concepts are currently applied in the consumer electronics industry, including personal computers and peripherals, is presented in Figure 4. Because technologies are improving so rapidly, product life cycles in this industry are extremely short, and a "reverse logistics" infrastructure has evolved to handle the large stream of obsolete hardware, recovering valuable components and in many cases refurbishing them for resale. The same overall system design could apply to most types of assembled durable goods, such as home appliances or automobiles.

Extending these concepts, William McDonough and Michael Braungart have given a particularly lucid and creative presentation of what they call an "eco-effective" system, one that eliminates waste by discovering benign or even restorative flows of materials and energy (*15*). This concept is illustrated in Figure 5. Any waste products from the product supply chain are converted into either biological nutrients (i.e., returned to the earth) or technical nutrients (i.e., recycled into the same or other product value chains).

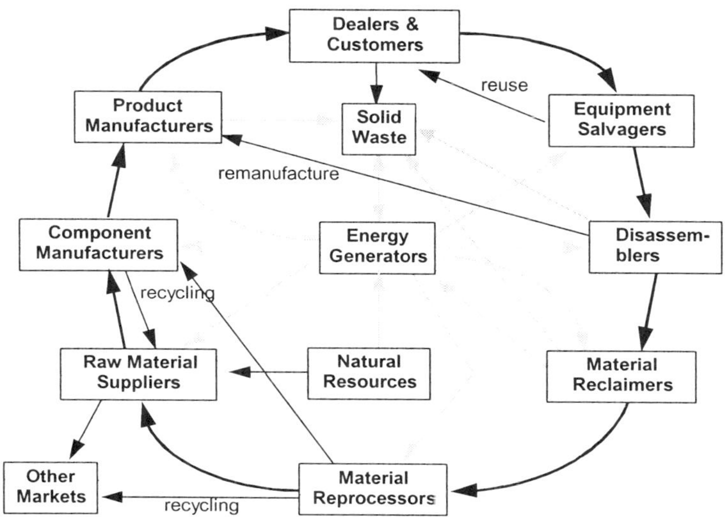

Figure 4. Industrial ecology model for consumer electronics.

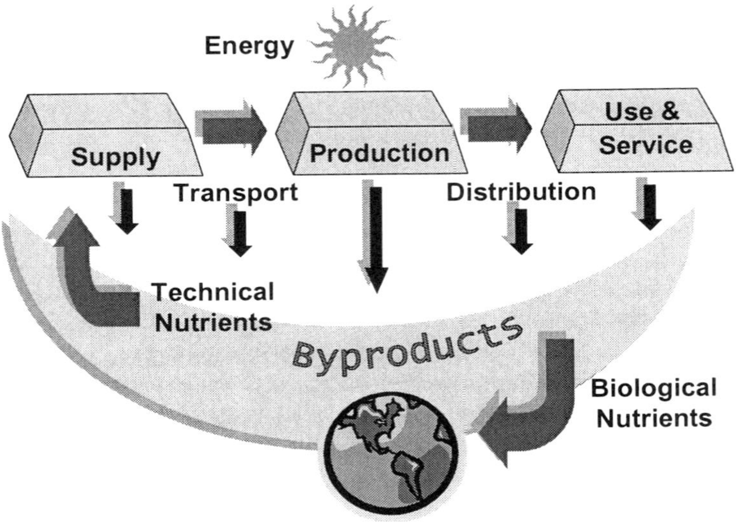

Figure 5. Flows of materials and energy in the biosphere.

Industrial Ecology in Practice

A much-publicized example of IE in practice is the industrial complex at Kalundborg, Denmark, where "symbiotic" relationships have developed among several large manufacturing facilities and a host of smaller ones (*16*). The main partners are a coal-fired power plant, a refinery, a plasterboard factory, and a pharmaceutical and enzyme plant operated by Novo Nordisk. IE flows include capture of waste heat for municipal uses and aquaculture, emission recovery (e.g., fly ash, sulfur) for substitute fuel and raw materials, and agricultural applications of waste streams. These relationships developed spontaneously based on economic incentives, but despite much effort at designing "eco-industrial parks," comparable examples are scarce.

The major impact of IE thinking has been to provide a fundamental design logic that guides awareness of environmental and social considerations in both the creative phase and the evaluative phase of new product or process development. Many companies have made efforts to systematize IE principles in the form of design guidelines. For example, companies as diverse as 3M,

Caterpillar, AT&T, Johnson and Johnson, and Bristol-Myers Squibb have developed Product Life Cycle Management programs that include "design for environment" (DfE) or "design for sustainability" guidelines. These programs include both review and enhancement of existing products and modification of the new product realization process to incorporate sustainability considerations.

The influence of IE thinking is also evident in recent actions of the U.S. Federal government. For example, the Environmentally Preferable Purchasing guidelines published by the U.S. Environmental Protection Agency specify products and materials that are considered environmentally friendly because of their recyclability, use of recycled content, and energy efficiency. Similarly, Executive Order 13123, "Greening the Government through Efficient Energy Management," directs Federal agencies to apply the principles of sustainable design and development to the siting, design and construction of new facilities, and it also directs agencies to optimize life-cycle costs, pollution, and other environmental and energy costs associated with the construction, life-cycle operation, and decommissioning of facilities.

There are numerous examples in various industries of companies that have realized financial and competitive advantages through the use of IE strategies. The efforts of leading chemical companies such as Dow and DuPont to develop sustainable products and to reduce their environmental footprints are well documented. Even in industries with a more traditional image, IE has shown great promise. The following are outstanding examples:

- Chapparal Steel in Midlothian, Texas, has developed a steel mill based on industrial ecology concepts. They formed a new company to recycle electric-arc furnace slag into cement production, reducing energy use and CO_2 emissions, and they constructed a world-class shredding and recycling facility adjacent to the plant. To enable these innovations, they acquired an exclusive separation technology for non-chlorinated plastics and are participating in byproduct synergy exploration with other firms (*17*).

- Collins & Aikman Floorcoverings in Dalton, Georgia, has been a pioneer in the use of innovative, sustainable technologies. They developed an adhesive-free carpet installation system, with 100% closed-loop recycled backing. As a result of product and manufacturing innovations, between 1993 and 1997 they were able to achieve a 68% increase in production volume, a 78% reduction in waste per square yard, 43% lower energy use, and 41% lower water use.

Green Chemistry as an Industrial Ecology Strategy

Green Chemistry, also known as "sustainable chemistry," is a particular branch of DfE that is concerned with environmentally benign chemical synthesis and processing. It differs from conventional DfE efforts in that DfE tends to focus on the mechanical design of products or packaging, resulting in incremental improvements for discrete products, (e.g., an inkjet printer or a bottle of shampoo). In contrast, green chemistry focuses on fundamental breakthroughs in product or process chemistry, deals with properties of materials at the molecular level, and, therefore, has the potential to discover more radical and more powerful solutions with greater beneficial impacts (*18*).

Green chemistry research can develop innovative ways to address many of the goals associated with IE (see, for example, (*19*, *20*)). Examples include:

- Reducing material intensity through the use of alternative chemistries that have higher "atom efficiency."
- Reducing energy intensity through lowering the energy requirements of endothermic reactions, for example, by using novel catalysis methods.
- Reducing the dispersion of toxic substances through discovery of "benign synthesis" pathways, or through elimination of organic solvents.
- Enhancing material recyclability by developing biodegradable materials or by finding methods to regenerate feedstocks from waste byproducts.
- Enhancing process yield and reducing waste by developing more effective separation and extraction methods, enabling byproduct recovery.
- Improving the durability and performance of products by developing novel materials or multilayered composites with customized physical properties.
- Maximizing the sustainable use of renewable resources by developing bio-based processing methods for converting biomass (e.g., soybeans, corn, cellulose) into useful materials.
- Developing innovative products that are environmentally benign (e.g., Rohm and Haas developed an eco-friendly marine antifoulant).

As an example of green chemistry innovations, researchers at Battelle Memorial Institute have been developing economically attractive polymer products based on renewable feedstocks derived from plant sources such as soybeans. Advances in plant genomics and genetic engineering have opened practical ways to build chemical functionality into the proteins and oils typically found in plant seeds, offering new paths for developing commercially viable products. This research program has resulted in the following developments:

- An environmentally compatible, soybean oil-derived plasticizer for use in the processing of polyvinyl chloride resin that replaces phthalate, benzoate, and other petroleum-derived products.
- A soy-based wood adhesive that performs better than current adhesives, is less costly, and will reduce emissions of volatile organic compounds.
- A novel, readily de-inkable toner technology based on soy resins, addressing a growing need in the manufacture of secondary fiber from office waste.
- Lactide polymer production from fermentation processes for use in the manufacture of biodegradable polymers.
- A cost effective "green" solvent, consisting of a water-based, biodegradable micro-emulsion, which uses the excellent solvent characteristics of the methyl ester of soybean oil.

While much of green chemistry is focused upon the development of new chemical pathways, there is also an important, related field of DfE in the area of green chemical engineering (see, for example, (*21*)). The main focus of this field is to improve the economic and environmental performance of industrial processes through optimization of process design and operating conditions. Figure 6 illustrates a heuristic design strategy based on a tree-structured logic, which can be used to reduce byproduct formation in continuous reaction processes (*22*).

Conclusions

The new-found emphasis on sustainable development within the business community signals an emerging synthesis between traditional business values and the concepts of environmental and social responsibility. Practically speaking, as global environmental and economic pressures intensify, the need for sustainability awareness will become a business imperative. The world population has surpassed six billion, while concerns about climate, water, land, and habitat preservation continue to mount. Rapidly developing economies around the world are creating growing markets for goods and services. These conditions are generating opportunities for companies to fundamentally change how they engage suppliers, operate facilities, and service customers.

In addition to new technologies, new production methods, and new management systems, companies that adopt SD will need a new, holistic paradigm to support creative research and innovation. Industrial ecology offers such a paradigm, at least for the environmental and economic aspects of sustainability. Specifically, green chemistry focuses on fundamental changes in material properties and process chemistries that can enable radical reductions in resource consumption and waste generation while creating economic value across the supply chain. The next logical challenge, as yet not addressed, is to develop a unifying conceptual paradigm that encompasses not only the environmental and economic aspects, but also the social aspects of sustainable development—a common framework for both natural and human capital.

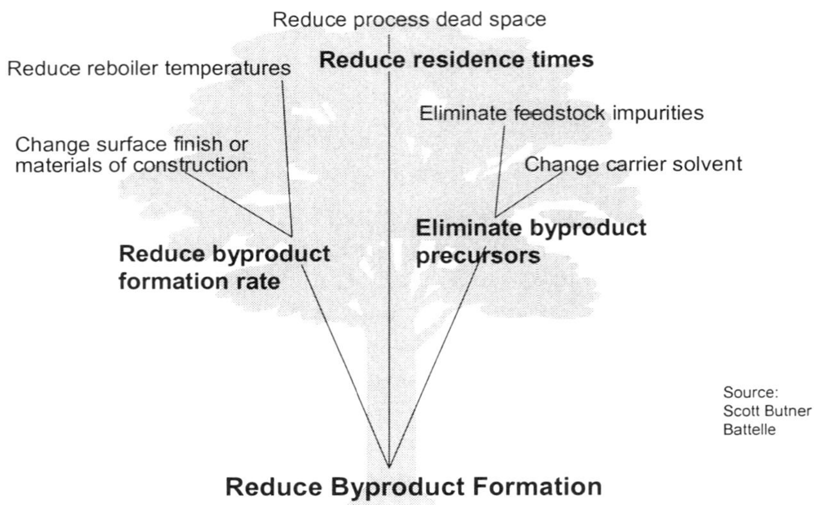

Figure 6. Heuristic process design rules.

References

1. World Commission on Environment and Development. *Our Common Future;* Oxford University Press: New York, 1987.
2. *World Development Indicators 2001*; World Bank: Washington, DC, 2001.
3. Keating, M. *Agenda for Change;* Centre for Our Common Future: Geneva, Switzerland, 1993.
4. Schmidheiny, S. with the Business Council on Sustainable Development. *Changing Course: A Global Business Perspective on Development and the Environment;* MIT Press: Cambridge, MA, 1992.
5. Fiksel, J.; Fiksel, D. G. From Here to Sustainability: A Global Perspective on Sustainable Development. *Chemistry Business*, April 2001.

6. Fiksel, J. Measuring Sustainability in Eco-Design. In *Sustainable Solutions: Developing Products and Services for the Future*; Charter, M., Tischner, U., Eds.; Greenleaf Publishing: Surrey, U.K., 2000, Chapter 9.
7. Global Reporting Initiative. *Sustainability Reporting Guidelines on Economic, Environmental, and Social Performance*; Global Reporting Initiative: Boston, June 2000.
8. Ayres, R. U. Industrial Metabolism. In *Technology and Environment;* National Academy of Engineering: Washington, DC, 1989, pp 23–49.
9. White, R. M. Preface. In *The Greening of Industrial Ecosystems,* Allenby, B. R., Richards, D. J., Eds.; National Academy of Engineering: Washington, DC, 1994.
10. Chiddick, D. Testimony before Senate Environment and Public Works Committee; February 23, 1993.
11. Tibbs, H. Industrial Ecology—An Agenda for Environmental Management. *Pollution Prevention Review,* Spring 1992.
12. Lowe, E. Industrial Ecology: A Context for Design and Decision. In *Design for Environment: Creating Eco-Efficient Products and Processes;* Fiksel, J., Ed.; McGraw-Hill: New York, 1996.
13. Robèrt, K. H. *The Natural Step: A Framework for Achieving Sustainability in Our Organizations*; Pegasus: Cambridge, MA, 1997.
14. Benyus, J. *Biomimicry;* William and Morrow: New York, 1997.
15. McDonough. W.; Braungart, M. The NEXT Industrial Revolution; *Atlantic Monthly* **1998,** *282* (4), pp 82–92.
16. Stigliani, W. M.; Anderberg, S. Industrial Metabolism and the Rhine Basin. *Options* (IIASA: Austria) **1991** (Sep).
17. Business Council for Sustainable Development–Gulf of Mexico. *By-product Synergy: A Strategy for Sustainable Development*; Radian International LLC: Austin, TX, 1997, p 4.
18. Anastas, P.; Warner, J. *Green Chemistry: Theory and Practice*. Oxford University Press: New York, 1998.
19. Anastas, P. T.; Lankey, R. L. Life Cycle Assessment and Green Chemistry: The Yin and Yang of Industrial Ecology. *Green Chemistry Journal* **2000,** (6), 289–295.
20. *Green Chemical Syntheses and Processes;* Anastas, P. T., Heine, L. G., Williamson, T. C., Eds.; ACS Symposium Series 757, American Chemical Society: Washington, DC, 2000.
21. *Green Engineering;* Anastas, P. T., Heine, L. G., Williamson, T. C., Eds.; ACS Symposium Series 766, American Chemical Society: Washington, DC, 2001.
22. Butner, R. S. A Structured Heuristic Approach to Integrating Pollution Prevention in Process Design. *Inst. of Chemical Engineers Environmental Protection Bulletin* #047; Rugby, U.K., March 1997.

Chapter 3

Research and Development Needs for a Fully Sustainable Biocommodity Industry

Charles E. Wyman

Thayer School of Engineering, Dartmouth College, 8000 Cummings Hall, Hanover, NH 03755

Lignocellulosic biomass provides a low-cost, uniquely sustainable resource for large-scale production of organic fuels and chemicals that can have low environmental impact while providing important economic and strategic advantages. Conversion costs must be competitive for these benefits to be realized, and biological processing routes can take advantage of emerging tools in modern biotechnology to dramatically improve economics. For example, significant advances have been made in biological conversion of lignocellulosic biomass into ethanol, and the private sector is working to apply the technology commercially. Raising the substantial capital needed is challenging for first-of-a-kind commodity projects, but understanding the fundamental principles governing bioprocessing steps will accelerate the rate of application. In addition, improved knowledge of cellulose and hemicellulose hydrolysis grounded in fundamental principles will facilitate improvements that are vital to the cost reductions needed to achieve competitiveness in an open market and to develop a range of sustainable biocommodity products. Premature attempts to force technological applications without building technical and economic foundations will be inefficient in resource allocation and will often fail.

Introduction

Lignocellulosic biomass, hereafter referred to as "biomass", includes agricultural (e.g., corn stover) and forestry residues (e.g., pulp and paper sludge), a substantial fraction of municipal solid waste (e.g., waste paper and yard waste), and ultimately herbaceous (e.g., switchgrass) and woody (e.g., poplar) crops grown to support large-scale uses (*1*). In this paper, we show that such low-cost raw materials are competitive in price with petroleum and can be biologically converted into a wide range of organic fuels and chemicals at a scale that can have tremendous environmental and economic benefits. Furthermore, biomass is the only known sustainable resource for production of such products that are so important to our lifestyle. Thus, while the sustainability of producing and using fuels and chemicals derived from fossil sources may be improved in terms of environmental impact, biomass provides the only resource for making such products that can also offer sustainability in terms of resource availability.

Although biomass will undoubtedly be the only option for making organic fuels and chemicals that can minimize greenhouse gas release and reduce our vulnerable dependence on petroleum, their cost must be competitive with petroleum-derived products to realize widespread use now. Given that petroleum refining has matured from limited production of lamp oil to large-scale manufacture of a broad spectrum of products with over a century of experience, it is important to be realistic about the magnitude of this challenge (*2*). Nonetheless, biological processing of biomass has capitalized on the powerful emerging tools of modern biotechnology to make substantial progress in this direction despite limited research funding for such a low-value product (*3*), and significant opportunities have been identified to further reduce bioprocessing costs to compete with conventional products (*4*,*5*). Consequently, the technology is competitive now in some applications, and further cost reductions are imminently possible.

Premature attempts to commercialize bioprocesses will fail without the political will to provide consistent support for incentives or other market measures or the ability to select and adequately finance projects with economic merit. Of course, a crisis atmosphere could arise that produces unrealistic quests for instant solutions to issues such as severe climate change, cartel reductions in petroleum availability, or other festering problems that receive little attention now. However, because a hasty response will fail without a proper foundation, patience is needed to develop a solid technology base that will support low risk scale up of cost-effective technologies that can transform biomass into a range of high volume commodity products and propel our society into a new era of resource and environmental sustainability with or without crises. We will attempt to present the background and a framework for such a metamorphosis.

The Biomass Resource and Refining to Biocommodities

Biomass has been shown to provide an abundant resource that would be more than adequate to produce quantities of organic chemicals equivalent to total current use (*6*). Although subject to more speculation, biomass could also potentially make a major impact on transportation fuels (*7*). As shown in Figure 1, biomass costing $42/ton is equivalent to petroleum at about $6.25/barrel on a weight basis, or approximately $13/barrel in terms of equivalent energy content (*3*, *8*). Because biomass from waste sources or from improved energy crops is expected to be available at costs less than this, its cost should not be a major impediment to the manufacture of fuels and commodity chemicals. Rather, the primary challenges are to insure a match between biomass supply and demand and to develop low cost processes that are competitive. The emerging fields of biotechnology and bioprocessing provide the powerful tools needed to reach such a goal.

Cellulosic biomass is made up of about 40 to 50% cellulose, 25 to 30% hemicellulose, 10 to 18% lignin, and lesser amounts of minerals, solubles, protein, and other components (*1*). Acids or enzymes can hydrolyze cellulose to glucose sugar while five sugars (arabinose, galactose, glucose, mannose, and xylose) are usually produced when acids or enzymes act on hemicellulose. As illustrated in Figure 2, these sugars can be fermented to various products such as ethanol and other alcohols, organic acids, glycols, and aldehydes, or the fermentation products or sugars themselves can be chemically transformed into products, opening up more possibilities (*3*, *9*). Lignin can be chemically converted into phenolics, aromatics, dibasic acids, and olefins or can be burned for heat and electricity for the conversion process, for export, or for both. Although not typically considered now, the high protein content in many types of biomass could be recovered for food or feed applications, extending the impact of biomass use.

The spectrum of biorefinery products made possible by the diverse structure of biomass can provide synergies analogous to modern petroleum and grain processing operations (*2*, *6*). Manufacture of fuels such as ethanol will support large-scale operations that achieve economies of scale, lowering costs as long as additional feedstock transportation costs are less than the incremental savings due to nonlinear scale up of capital and some operating costs. Portions of the resulting low cost sugars can be processed into lower volume, higher value chemicals and materials, potentially increasing profitability if the margins are greater. In addition, excess heat or power derived by burning lignin can be sold

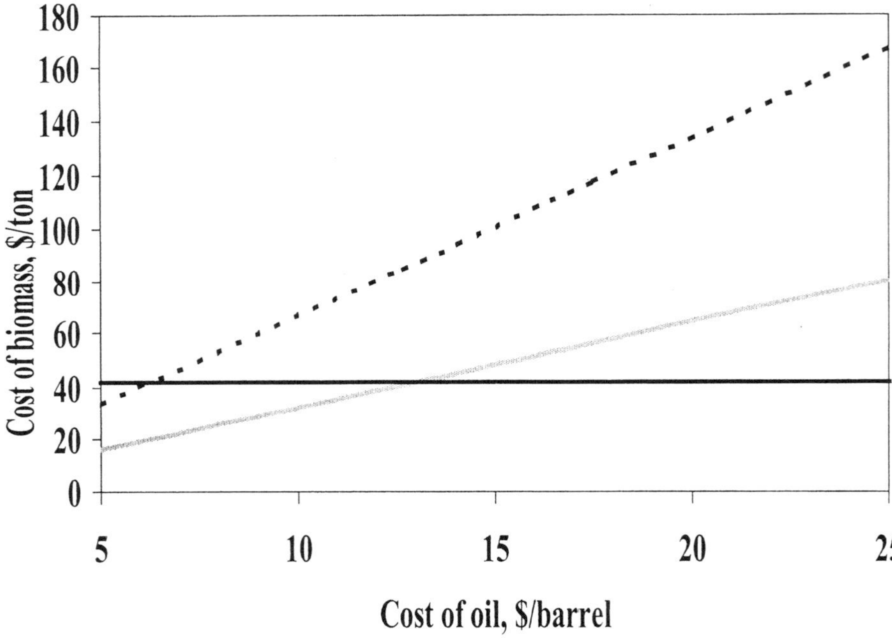

Figure 1. Cost of biomass versus cost of petroleum based on mass (dashed lines) or energy content (solid line), with the horizontal line indicating biomass costing $42/ton (3, based on reference 8).

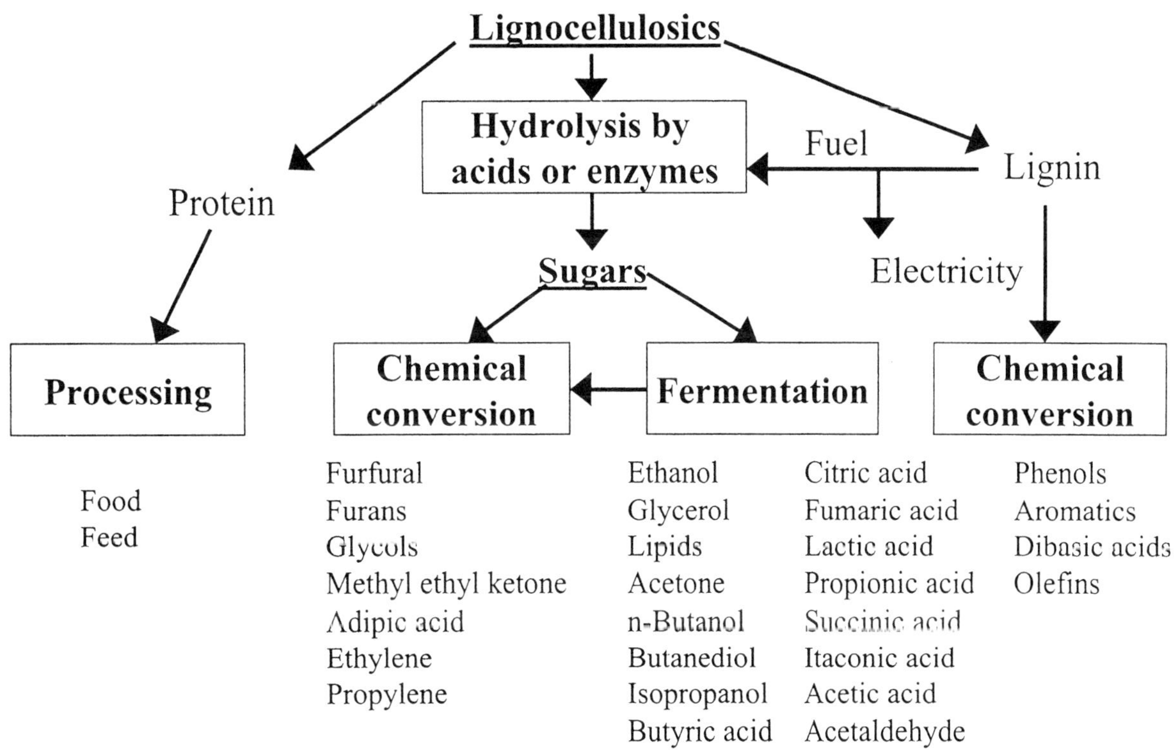

Figure 2. A biorefinery concept for conversion of cellulose, hemicellulose, lignin, and protein in biomass into fuels, chemicals, materials, heat, power, food, and feed (3, based on reference 9).

at a lower price than possible in a dedicated power plant, allowing biomass power to compete in deregulated utility markets. Biological processing of biomass to commodity products through such an approach has recently been termed "biocommodity engineering" to distinguish the unique features of this emerging discipline from more typical current applications of biotechnology to high-value, low-volume health care products (*8*).

Sustainability and Other Benefits of Biocommodity Products

Sustainable development can be defined as meeting the needs of the present without compromising the ability of future generations to meet their needs (*10*). Thus, petroleum is not a sustainable resource because it is finite and contributes carbon dioxide and other greenhouse gases that could lead to global climate change. Furthermore, it can be argued convincingly that burning of fossil resources as fuel squanders a valuable resource for making chemicals by future generations, and although the environmental consequences of fossil resource use can be improved for some chemicals, fossil-based technologies cannot address overall sustainability. On the other hand, biomass offers a unique sustainable resource from both a supply and impact perspective when low impact crops are used, fossil resources are not required during biomass production and processing, and environmentally benign products are made (*5*, *11*).

By definition, all organic products are ultimately derived from biomass, with petroleum being a depletable biomass reservoir. Thus, biomass is the only sustainable resource for making organic products, as well as the only source of human food either directly as fruits, vegetables, and grains or indirectly as animal feed. On the other hand, heat, mechanical energy, electricity, and hydrogen could be derived by many other sustainable technologies such as wind, hydropower, ocean thermal energy, photovoltaics, and nuclear power. This unique match of biomass to providing food, feed, and organics coupled with the large demand for these products and limitations in biomass availability suggests that priority for biomass utilization in a rational world should be given to production of food, feed, and organic chemicals and fuels, in that order.

Some clarification is worthwhile about the sustainable use of biomass. Several studies have shown that very little fossil energy is needed to plant, grow, harvest, and transport biomass and ship final products to their markets. Furthermore, burning the lignin in the feedstock can provide more than enough heat and electricity to process biomass to ethanol (*7*, *12–14*). Thus, the ratio of energy available from bioethanol to fossil energy inputs is significant. However, because far more energy is contained in ethanol than is required to make it, bioethanol itself could be used to fuel vehicles for planting, harvesting, and transportation operations while other inputs such as fertilizer could be derived from sustainable sources instead of the fossil sources commonly assumed in such studies. This is analogous to the fact that tankers burn petroleum to cross the oceans and still deliver much more than they consume at a low cost. Thus, provided the crops employed have little impact on soils, biomass-derived products such as ethanol are renewable.

While bioethanol clearly can be made with little, if any, fossil fuel inputs, the energy balance for some polymeric products has been shown to be less favorable than for fossil fuel options (*15*, *16*). In this case, aerobic fermentations were employed to make viscous materials that require far more energy input than the anaerobic technologies applied for lower viscosity ethanol manufacture, and energy intensive separations were needed for product recovery. In addition, fermentation sugars were derived from corn, a feedstock that requires more fuel and fertilizer to produce than biomass, and fossil sources are used to power the process. On the other hand, use of biomass sugars for fermentation and lignin to generate power in the process would reduce fossil fuel use and improve sustainability, although additional biomass inputs could be needed to meet the high-energy demands of this technology. This study illustrates that it can be misleading to assume that all biomass-based technologies are sustainable and that careful analysis is needed to insure that the process is properly configured and uses an appropriate raw material. These results also support the idea that petroleum should be saved to manufacture products for which it is so uniquely suited and not squandered on uses for which there are sustainable alternatives.

Biomass conversion offers many other important benefits in addition to sustainability. These include disposal of solid organic wastes, reduced trade deficits, improved domestic security, creation of rural agricultural and manufacturing employment, and better air quality. However, because these attributes have been described extensively elsewhere, the reader is referred to such sources (*1*, *5*, *7*, *8*, *11*).

Research Progress

In the early 1980s, evaluations of biomass conversion to ethanol were conducted to discover a quick fix that would ameliorate the high price of petroleum (*3*). Although biomass conversion was found to not yet be competitive with making ethanol from corn, primarily because the mixture of five sugars released in hemicellulose hydrolysis could not be converted to ethanol at high yields or used for other products, these studies were used to develop a common basis for comparison of bioethanol technologies and to define research priorities (*17*). These results, coupled with a policy shift to funding long-term, high-risk research, led to targeting enzymatic hydrolysis of cellulose to glucose because high yields and competitive costs were feasible through the application of modern biotechnology. Although directed at bioethanol, all of the key steps through hemicellulose and cellulose hydrolysis are applicable to the production of biomass sugars ("biosugars") that can be used to make many other products.

Technoeconomic evaluation models have been updated periodically to track steady research progress in lowering the cost of bioethanol from about $4.60/gal in the 1980 timeframe to a value competitive with ethanol from corn now (*5*, *14*, *17*). However, these cost projections are measures of research progress based on a number of needed assumptions and simplifications, and processing costs will be sensitive to variabilities in feedstock costs, project financing approach, technology used, and other factors that require more detailed estimates. Nonetheless, they show that bioethanol can be competitive.

Two advances were instrumental to reducing the cost of bioethanol by about a factor of four over the past 20 years (*5*). First, improvements were made in our ability to make biosugars from cellulose and hemicellulose by "overcoming the recalcitrance of biomass". These advances were achieved through optimizing dilute acid biomass pretreatment to improve sugar yields from hemicellulose and to increase the recoverability of glucose in subsequent cellulose digestion by enzymes. In addition, better cellulase enzymes that improve glucose release and reduce the use of expensive enzymes along with process integration to improve rates, yields, and concentrations were important.

A second improvement category, "overcoming the diversity of biomass sugars" by genetically modified organisms, was critical to realizing competitive costs. Prior to this, ethanol yields from all five biomass sugars were inadequate, disposal costs of unused sugars were too high to be competitive, and markets for alternative products from these sugars were not sufficient to support large-scale bioethanol manufacture regardless of other improvements. But now a new class of genetically modified organisms makes it possible to derive revenue from all sugars, with the importance of this breakthrough recognized by the awarding of U.S. Patent 5,000,000 (*18*, *19*). Conversion of biosugars to products other than ethanol will also require that high product yields be achieved from all five sugars in biomass, although unused sugars could be converted to bioethanol.

Challenges in Commercializing Biosugar Technologies

Attempts to apply bioethanol technology before it was possible to ferment all sugars at high yields were doomed, because revenues were insufficient to cover costs. However, with the improvements discussed, the cost of making bioethanol and many other products from biosugars is now projected to be competitive, and firms are pursuing market applications. Although some express frustration that commercial processes are not yet in operation, it is important to realize that significant hurdles must still be overcome. Unfortunately, promising technology is not sufficient in itself, and the steps to commercializing commodity products in particular are demanding and slow.

Because economies of scale are important for most commodity processes, biorefineries must be large, and total capital costs become very high. Thus, large sums of capital must be raised, a very demanding task, from three typical sources of support: venture capital, equity, or project. Venture capital often expects a higher rate of return on investment than is possible for new commodity products. Internal equity financing is possible for large companies, but hurdle rates can be too high to accommodate contingencies and other margins for first-of-a-kind technology. Project financing offers lower rates of return for debt holders, but the demands on project certainty and longevity are significant.

In any financial arrangement, uncertainty and risk for biocommodity projects must be minimized (*20*). This can require that contracts be in place for an adequate supply of biomass to the plant over much of its financial life. Often, contracts are needed for all off-take over the project life. In addition, the feedstock quality typically must be demonstrated to satisfy process needs throughout the operating year, including the effects of storage, and product and coproduct quality must meet market specifications. Finally, the engineers and contractors responsible for engineering, procurement, and construction may be required to provide process guarantees for yield and quality.

Meeting these and many other requirements to mitigate uncertainty and risk is demanding for a first-of-a-kind project. Long term feedstock and off-take agreements result in higher costs for raw materials and lower product selling prices to make these arrangements attractive, squeezing profit margins. In addition, those accepting risk through process guarantees take measures to protect themselves from losses. Such firms would prefer to scale up by no more than a factor of 10 from one scale to the next, but successive scale up strategies of this type are too expensive and risky for most operators to afford. Thus, in addition to normal contingencies to cover unexpected occurrences such as bad weather, a contingency will also be applied in case the process does not perform as well at a large scale as it did in smaller systems, and extra equipment is often included to insure that poorly understood streams can be processed. Technologies and equipment with established commercial performance will also be employed wherever possible to reduce risk even though such equipment may be more costly or less effective than more advanced but less practiced options. Beyond that, other contingencies are applied for new technologies to compensate for the unknowns that often occur with new technologies.

The overall effect of all of these contingencies and safety factors is to drive the capital costs far higher and to squeeze profit margins more than is apparent from costing the core technology alone (*4*). We could view the cost structure as a series of concentric rings, with the center circle representing the cost of the technology itself as pictured for a fully mature process. Surrounding that circle would be a ring representing normal contingencies typical of any project The sum of these two circles is what we would expect to pay based on the process concept and laboratory performance data. However, additional cost rings need to be included to account for performance margins, extra equipment, and contingencies for the new technology unknowns discussed above, adding significantly to the cost. As a result, the total capital cost of the new technology will undoubtedly be far more than anticipated in the early stages or than one would expect once performance and equipment are well established as the same technology matures. Contractual requirements for feedstock and off take squeeze the profit margin, reducing returns on capital. Unfortunately, the difference between the true technology costs and those including risk coverage can be so great as to stop commercialization despite the substantial promise for the long term. The cost gap that develops as new technology moves from the laboratory bench to commercial use is often called the "Valley of Death" and can terminate its use despite tremendous promise and low-cost potential.

The nature of commodity products presents an additional challenge for market entry (*6*). The good news is that a large market has been established that one can sell into as long as price and performance requirements are met. However, it is difficult to establish a competitive advantage beyond price, and those already in the market will likely have the upper hand in terms of equipment that is partially or fully (cash cow) paid-for. Thus, financiers will often expect that the total product cost including cash costs and capital amortization for new entries must be less than the cash costs for existing technologies to insure survival in a price war. It can also be required that the margin between revenue and cash costs must be more than is needed to cover debt service, adding even more to the challenges of financing the project. Finally, first-of-a-kind projects have no reference point for benchmarking, reducing the comfort level for many financial institutions.

Some measures can be taken to help compensate for uncertainty and risk associated with application of new biocommodity technologies (*21*). First, waste biomass can be used at little if any cost to the project. Additionally, it may be possible to integrate the facility into an existing process such as a biomass power plant to take advantage of some existing equipment, reducing capital costs, as well as in-place infrastructure and permits. Use of low-cost debt funding, particularly as tax-free government bonds, can lower interest rates and capital amortization. Higher value coproducts such as chemicals from lignin could also improve revenues for initial operations. However, in spite of the advantages such measures can afford, it is still likely that government support will be essential to implement technology for the first time.

Research and Development Implications for Commercializing Biocommodity Products

Contingencies built into capital costs to meet process guarantees coupled with requirements for feedstock and product contracts and safe profit margins are major hurdles for new products. These requirements are even more challenging for new routes to commodities that must compete immediately with established technologies that have benefited from substantial learning curves. Although operation of integrated operations at a near-commercial scale can overcome many of these obstacles, capital and operating costs are so high and the risk so great that financing such operations is almost impossible, at best. However, improving our ability to predict performance of large-scale systems and further reductions in costs can be invaluable in overcoming these obstacles.

Improving the Accuracy of Process Scale Up

Commercial processes will not necessarily duplicate the laboratory systems often used to gather data at a scale many orders of magnitude smaller. For example, dilute acid hydrolysis of biomass may be performed manually in a simple batch laboratory operation. However, because finding suitable commercial equipment for solids handling is often challenging, the sequence of (1) adding acid to biomass, (2) mixing to insure uniform distribution, (3) allowing time for acid to fully penetrate pores, (4) pressurizing the resulting material to reaction conditions, (5) heating directly or indirectly, (6) holding the heated acidified biomass for a set reaction time, and (7) rapidly cooling the material to quench the reaction and prevent degradation of the sugars released, will be far more complex (*14*). The question then becomes how to predict the performance of commercial systems. For example, process engineers may need data on diffusivity of acids in biomass, solid/liquid mixing results, biomass heat transfer constants, and other design information typically not considered in conversion versus reaction time data gathered in a simplified laboratory setup. Armed with insufficient information, safety margins must be applied to insure that the project cannot fail, typically at considerable cost.

The importance of developing a fundamental understanding of technology to support its application has been described in the book *Pasteur's Quadrant* by Donald Stokes (*22*). The point is made that Vannevar Bush established a dichotomy between basic and applied science during World War II in which basic research is directed at understanding the scientific basis of phenomena without an application in mind, while applied research is viewed as using technology without trying to understand it. However, Stokes presents the case that research can be viewed in terms of a two-dimensional matrix that considers applications in one dimension and the quest for fundamental understanding in the other. In this context, pure basic research as classically defined would emphasize knowledge without concern for use, while pure applied research would favor use without concern for understanding. In general terms, Stokes favored consideration of both elements—applications and understanding—in a sector labeled "Pasteur's quadrant" in honor of the great scientist who laid the foundations of microbiology with a view to applications such as curing disease. In this mode, applications benefit from an improved fundamental understanding that illuminates cause-and-effect relationships.

Similarly, biomass research can benefit from an emphasis on developing fundamental principles that support commercial applications. Thus, for example, it is important to understand the impact of key process parameters on performance rather than simply gathering data on concentration versus time profiles for hydrolysis, fermentation, and other biomass operations. We need to predict such things as how fast acid or other hydrolysis chemicals diffuse through biomass, how such chemicals must be dispersed to penetrate the material, how long biomass must be soaked in such chemicals, how rapidly biomass can be heated by direct or indirect means, how recombinant organisms behave when grown in large-scale fermentors, how key nutrients must be provided at a large scale, and how shear rate affects performance in large scale fermentations. There are countless such questions that need to be addressed in a clear and convincing way to comfort engineers and investors financially responsible for projects costing tens to hundreds of millions of dollars. Failure to build confidence in such information will likely result in lost opportunities to apply technologies that have considerable promise for low-cost applications.

Advancing Biocommodity Technologies to Reduce Costs

Although cost projections show that bioethanol is presently competitive with ethanol from corn, incentives are required for ethanol from either feedstock to compete financially in the current fuel market. Furthermore, risk mitigation for the application of new technology for making ethanol or other new products from biomass can drive up costs and impede market entry, as discussed above. However, several studies have shown that additional cost reductions are feasible that would make bioethanol competitive as a pure fuel in the open market, drastically expanding the impact of the technology and its benefits (*4*, *5*). Many of the same advances are also applicable to other products that can be derived from biomass, thereby improving their competitiveness as well.

Three different approaches have been applied to show the low-cost potential of biomass: sensitivity studies built around existing economic analyses (*14*, *23*), allowable cost projections (*24*), and advanced technology scenarios

(*4*). Each shows that biomass can be converted into ethanol at costs that could compete in the open market, but because a summary of each has been presented previously (5), the former two will not be discussed here. However, a few key points are presented below about the advanced technology study because it is relevant to other products that could be made in a biorefinery.

The most expensive processing steps for converting biomass to ethanol or other products are pretreatment, which contributes about one-third of the cost, and biological conversion of cellulose to glucose and fermentation of glucose and other sugars to ethanol, collectively representing almost 40% of the total (*4*). Three technology scenarios were investigated to improve these costs. The first was built on a published base-case process that was believed to be achievable with existing technology. A second case, termed advanced technology, was defined to combine features of existing commercial products for other feedstocks and to represent technology judged to have the features most likely for a mature process, analogous to the evolution from early petroleum processing to modern refining technology. The third case, called best technology, envisioned the ultimate potential for research and development advancements for processing biomass, a state that provides an idea of the upper end of improvements.

To summarize, this analysis showed that attaining the advanced technology scenario could achieve costs of about $0.50/gal, a level that would compete with conventional fuels in the open market, while the best possible technology case could cost as little as $0.34/gal. The advanced technology pretreatment approach was patterned after some of the promising features of flow-through hot water only systems with high yields, short residence times, and use of little if any chemicals for hydrolysis. The advanced biological system also employed a consolidated bioprocessing operation that combined the operations of enzyme production, cellulose hydrolysis, and fermentation of all sugars in one step. Although other configurations could no doubt realize similar costs, the point was that advanced processing technologies, and not simple yield improvements alone, are essential to realizing competitive costs. By similar reasoning, it is expected that novel conversion configurations will also be essential to reducing the cost of making other products from biomass.

Development of advanced process configurations is challenging, and progress is slowing as we approach the final targets. However, advances to date have been largely by trial-and-error, and more efficient routes will accelerate the next steps. In particular, our knowledge of the mechanisms for many of biomass conversion systems is limited, and more emphasis on understanding the fundamental principles governing biomass conversion would provide a platform to help define and apply such advancements, reducing costs for both. This approach would be synergistic with supporting applications of current technologies.

Conclusions

In addition to traditional, vital uses for production of food and feed, biomass is the only sustainable resource for production of organic fuels and chemicals, and because many of these biomass products provide powerful environmental, economic, and strategic benefits, biomass provides a unique combination of sustainable benefits from both a resource and an environmental perspective. Thus, development and commercialization of new processing technologies that targets these important applications merits priority.

Hemicellulose and cellulose in biomass can be broken down to release sugars that can be fermented or chemically reacted into many commodity products. The lignin portion can be burned to provide heat and electricity to power the conversion process and for export, while biomass protein could be recovered for food and feed applications. The minerals in biomass could also be employed for commercial products. Tremendous progress has been made in reducing the cost of biological processing of biomass to ethanol and other commodity products, to the point where many are receiving commercial attention. However, significant challenges remain in applying such technologies due to their inherently high capital costs, the low value of commodity products, and uncertainty for first time commercial uses. Thus, only projects that are economically and technically viable will be successful, and improving our fundamental understanding of bioprocessing technology is vital to support low-cost scale up to commercial operations and to provide a foundation for technological improvements. In addition, more attention must be devoted to advancing technologies that promise to radically reduce the cost of pretreatment and biological processing to achieve competitiveness in an open market without subsidies, with applied fundamental research also being important to this quest.

Acknowledgments

Contribution of this paper was made possible through the support of the Thayer School of Engineering at Dartmouth College and the EPA/NSF Technology for a Sustainable Environment Program.

References

1. *Handbook on Bioethanol: Production and Utilization*; Wyman, C. E., Ed.; Applied Energy Technology Series; Taylor & Francis: Washington, DC, 1996; pp 1–424.
2. Yergin, D. *The Prize;* Simon and Schuster: New York, 1991.

3. Wyman, C. E. Twenty Years of Trials, Tribulations, and Research Progress in Bioethanol Technology: Selected Key Events along the Way," *Appl. Biochem. Biotech. 2001, in press.*
4. Lynd, L. R.; Elander, R. T.; Wyman, C. E. Likely Features and Costs of Mature Biomass Ethanol Technology. *Appl. Biochem. Biotech.* **1996,** *57/58*, 741–761.
5. Wyman, C. E. Biomass Ethanol: Technical Progress, Opportunities, and Commercial Challenges. *Ann. Rev. Ener. Envir.* **1999**. *24*, 189–226.
6. Wyman, C. E.; Goodman, B. J. Biotechnology for Production of Fuels, Chemicals, and Materials. *Appl. Biochem. Biotech.* **1993,** *39/40*, 41–59.
7. Lynd, L. R.; Cushman, J. H.; Nichols, R. J.; Wyman, C. E. Fuel Ethanol from Cellulosic Biomass. *Science* **1991**, *251*, 1318–1323.
8. Lynd, L. R.; Wyman, C. E.; Gerngross, T. U. Biocommodity Engineering. *Biotech. Prog.* **1999,** *15*, 777–793.
9. Wyman, C. E. Biological Production of Chemicals from Renewable Feedstocks. Presented at the American Chemical Society National Meeting, Washington, DC, August 26, 1990; 200: 39-CELL Part 1.
10. Jacobs, M. Toward a sustainable future. *Chem. Eng. News.* March 6, 2000, p 5.
11 Lynd, L. R. Overview and Evaluation of Fuel Ethanol from Cellulosic Biomass: Technology, Economics, the Environment, and Policy. *Ann. Rev. Ener. Envir.* **1996,** *21*, 403–465.
12. Wyman, C. E. Alternative Fuels from Biomass and Their Impact on Carbon Dioxide Accumulation. *Appl. Biochem. Biotech.* **1994,** *45/46*, 897–915.
13. Tyson, K. S. *Fuel Cycle Evaluations of Biomass-Ethanol and Reformulated Gasoline*; NREL/TP-463-4950, DE94000227; National Renewable Energy Laboratory: Golden, CO, 1993; Vol. I.
14. Wooley, R.; Ruth, M.; Glassner, D.; Sheehan, J. Process Design and Costing of Bioethanol Technology: A Tool for Determining the Status and Direction of Research and Development. *Biotech. Prog.* **1999,** *15*, 794–803.
15. Gerngross, T. U. Can Biotechnology Move Us Toward a Sustainable Society. *Nat. Biotech.* **1999,** *17*, 541–544.
16. Gerngross, T. U.; Slater S. C. How Green are Green Plastics. *Scien. Amer.* **2000,** *283* (2), 36–41.
17. Wright, J. D. Ethanol from Biomass by Enzymatic Hydrolysis. *Chem. Eng. Progr.*, August 1988, 62–74.
18. Ingram, L. O.; Conway, T.; Alterthum, F. Ethanol Production by *Escherichia coli* Strains Co-Expressing Zymomonas PDC and ADH Genes. U.S. Patent 5,000,000, March 19, 1991.
19. Ingram, L. O.; Conway, T.; Clark, D. P.; Sewell, G. W.; Preston, J. F. Genetic Engineering of Ethanol Production in *Escherichia coli*. *Appl. Envir. Micro.* **1987**, *53*, 2420–2425.

20. Keller, J. B.; Plath, P. B. Financing Biotechnology Projects: Lender Due Diligence Requirements and the Role of Independent Technical Consultants. *Appl. Biochem. Biotech.* **1999**, *77–79*, 641–648.
21. Wyman, C. E.; Goodman, B. J. Near Term Application of Biotechnology to Fuel Ethanol Production from Lignocellulosic Biomass. In *Opportunities for Innovation in Biotechnology*; Busche, R. M., Ed.; NIST GCR 93-633; National Institute of Standards and Technology: Gaithersburg, MD, 1993; pp 151–190.
22. Stokes D. E. *Pasteur's Quadrant: Basic Science and Technological Innovation*; Brookings Institution Press: Washington, DC, 1997.
23. Hinman, N. D.; Schell, D. J.; Riley, C. J.; Bergeron, P. W.; Walter, P. J. Preliminary Estimate of the Cost of Ethanol Production for SSF Technology. *Appl. Biochem. Biotech.* **1992**, *34/35*, 639–649.
24. Wyman, C. E. Economic Fundamentals of Ethanol Production from Lignocellulosic Biomass. In *Enzymatic Degradation of Insoluble Carbohydrates*; Saddler, J. N., Penner, M. H., Eds.; ACS Symposium Series 618; American Chemical Society: Washington, DC, 1995; pp 272–290.

Chapter 4

Tetraamido Macrocyclic Ligand Catalytic Oxidant Activators in the Pulp and Paper Industry

Terrence J. Collins[1], Colin P. Horwitz[1], Alexander D. Ryabov[1], Leonard D. Vuocolo[1], Sayam S. Gupta[1], Anindya Ghosh[1], Nadine L. Fattaleh[1], Yelda Hangun[1], Bradley Steinhoff[1], Christopher A. Noser[1], Evan Beach[1], Duane Prasuhn, Jr.[1], Trevor Stuthridge[3], Kathryn G. Wingate[3], Jenny Hall[2], L. James Wright[2], Ian Suckling[3], and Robert W. Allison[3]

[1]Department of Chemistry, Carnegie Mellon University, 4400 Fifth Avenue, Pittsburgh, PA 15213–2683
[2]Department of Chemistry, University of Auckland , Private Bag 92019, Auckland, New Zealand
[3]NZ Forest Research Institute, Private Bag 3020, Rotorua, New Zealand

One-time-use small molecule catalytic activation of H_2O_2 promises valuable benefits for the pulp and paper (P&P) industry. Tetraamido macrocyclic ligand (TAML) catalysts activate H_2O_2 to rapidly bleach wood pulp at high selectivity from room temperature to 90 °C, providing a new totally chlorine free (TCF) approach. In addition, TAML-activated peroxide bleaches "color" in ClO_2 bleach plant effluents using low peroxide and minute catalyst concentrations in minutes at room temperature, giving 50 to 90% color reduction depending on the effluent stream. Here we summarize potential TAML/peroxide applications that promise to improve capital and operating costs, product quality, environmental performance, and energy efficiency. TAML-activated peroxide is applicable in diverse fields of technology among which the P&P field is but one.

Introduction

We begin by reflecting on the lack of catalysis in the pulp and paper (P&P) industry (*1*). Throughout the 20th century in the chemical industry, metal-based small-molecule catalysts were the engines driving many of the most profitable and most widely practiced technologies. In contrast, metal-based catalysis held no presence in the P&P industry. Yet the principal business of the P&P industry is applied chemistry on a gigantic scale. So if the chemical industry can derive great commercial benefit from metal-based catalysis, why not the pulp and paper industry? Environmental factors and the vitality of pulp and paper science as a research field heighten the significance of this question, as does the great need to *green* applied chemistry across the board (*2*).

In the last decades, environmental concerns have forced changes in the stoichiometric chemistry that is the signature of pulp delignification and bleaching. This move has been driven especially by the desire to eliminate the production and distribution throughout the environment of the potently bioaccumulative, carcinogenic, and powerfully endocrine-disrupting (*3*) polychlorinated dibenzo-dioxins and -furans (PCDDs and PCDFs; those chlorinated in the 2,3,7,8-positions are the most toxic (*4*). These compounds are deadly toxins to humans (*5, 6*) as well as other mammals (*3, 4, 6*). Because incorrect information has been disseminated concerning the dangers that PCDDs and PCDFs represent to humans (*6*), it has become an important burden for chemists to help set the public record straight whenever the opportunity arises. Uncertainty still clouds our understanding of the full impact of endocrine-disrupting chemicals (EDCs), but it is understood that their effects include disrupting the body's natural control over the reproductive system by mimicking or blocking the regulatory functions of the steroid hormones or by altering the amounts of hormones in the body. Both because these persistent organochlorine pollutants have been shown to cause severe reproductive damage to certain mammals at miniscule body burdens and because comparable tests cannot be performed on humans, it is important that we reinvent chemical technology such that the sources of known anthropogenic EDCs are eliminated altogether (*3–7*).

The principal change adopted by the P&P industry to cope with PCDDs, PCDFs, and other chlorinated organic pollutants has been to shift to chlorine dioxide bleaching, called "D bleaching". Collectively, the chlorinated byproducts of pulp bleaching are called "AOX" for "absorbable organic halogen", because in one form of measurement, they are absorbed upon activated charcoal. The industrial change to D bleaching has been adopted because it produces much less chlorinated organics than chlorine bleaching, called "C bleaching" and this especially true for PCDDs and PCDFs (*8*). Totally chlorine free (TCF) approaches have not been adopted with the same enthusiasm, because D bleaching exhibits superior selectivity for attacking lignin over cellulosics in the pulp than any TCF alternative, thus leading to a higher quality product. However, while D bleaching now dominates the marketplace, it continues to rely on stoichiometric chemistry. D bleaching is capital intensive, and it is a credit to the P&P industry in many countries that so much of the required capital to move away from C bleaching worldwide has been spent because of the environmental concerns. Operating expenses of D bleaching are also high and are strongly tied to energy prices because the manufacture of ClO_2 is energy intensive.

Although D bleaching is environmentally much more sound than C bleaching, it is still not free of environmental negatives. Because chlorine dioxide reacts with chloride ion to give chlorine, trace quantities of persistent organochlorine pollutants remain *inescapable* components of the enormous effluent streams produced from D bleaching. Chlorine dioxide may also produce some of these pollutants itself. What distinguishes the toxicity of chlorine-based bleaching from TCF bleaching is the persistence of organochlorine pollutants (*6*). TCF processes may produce effluent that is more toxic to aquatic life in the short term, but the toxins are biodegradable. This argument of the importance of persistence is often side-stepped in discussions of the relative toxicities of effluent streams from the P&P industry.

The waste products of chlorine-based bleaching end up in natural water systems. From a global perspective, there is still much to be done in reducing dioxin production in the P&P industry, especially in less developed countries where chlorine is still employed because of capital and operating cost factors. Because of their persistence and their vapor pressures that allow for atmospheric distillation, dioxins represent not just a local threat, but a global threat, no matter where they are generated (*6*). Because of their potent bioaccumulative properties, extreme general toxicity, and especially their endocrine-disrupting properties, research and regulatory processes should proceed such that all anthropogenic sources of PCDDs and PCDFs finally can be eliminated as a step towards attaining a technology base that can sustain our race and its civilization (*2*). A highly effective TCF process that is less expensive to operate than D bleaching and not capital intensive could help significantly to extend the move away from dioxin-producing pulp bleaching.

AOX compounds are not the only environmental concern with chlorine-based bleaching. Effluent streams from D bleaching are contaminated with "color," the recalcitrant tea-colored stain that is a signature of the process. Color thus contaminates the millions of gallons of water expelled daily by the P&P industry into natural waterways.

Catalysis by its very nature has the potential to enable improvement in delignification and bleaching at greatly reduced capital expenditure compared to industry norms, while at the same time improving operating costs and environmental performance. Catalysis has not flourished in pulp bleaching because the requirement has been for reactive homogeneous oxidant activators, especially activators of oxygen or hydrogen peroxide, to enter the heterogeneous pulp substrate and selectively attack the lignin. Such catalysts have been largely absent from chemistry in general because the ligand systems needed to direct the active metal center to favor selective processes have themselves been incompatible with strongly oxidizing media. Thus, the principal problem thwarting researchers searching for effective systems has been ligand decay leading to short catalyst lifetimes.

For most purposes, in producing a catalyst for activating hydrogen peroxide, one must avoid Fenton chemistry, the chemistry that occurs when metal ions such as ferrous or ferric ions interact with peroxides, especially in water. This chemistry is characterized by the liberation of oxygen, noted when peroxide is used to disinfect wounds, coupled to the production of a flux of hydroxyl radicals. The hydroxyl radical is a potent H-atom abstractor because the O–H bond of water is particularly strong. At 119.6 $kcal/mol^{-1}$, it is stronger than most C–H bonds. Thus, Fenton chemistry is not selective enough to be useful for most applications and tends instead to bring about the oxidative decomposition of organic matter. This is useful only when decomposition is the goal, as in the disinfection of wounds.

One wants instead for a metal ion to produce with peroxide another type of reactive intermediate, one that is still very reactive but much more selective than the hydroxyl free radical itself. One such intermediate is a metal–monooxo complex. Others are peroxo complexes of various types. The oxo complex arises when the catalytic metal ion abstracts an oxygen atom from hydrogen peroxide and discards water, making a two-electron oxidized metal-oxo species, as is found in the iron-based cytochrome P-450 enzymes. We believe that to induce this reaction channel over Fenton chemistry, one must bind to the metal ion electron-releasing ligands that favor the formal metal-based two-electron oxidation that coincides with metal–oxo formation. We have demonstrated the value of this idea in manganese chemistry by using strongly σ-donating deprotonated secondary amides and alkoxides to produce the first manganese(V)-monooxo complexes that were stable enough to be isolated and structurally characterized (*9–12*).

Iron is abundant in the Earth's crust and is a common element in biochemistry. This makes iron an especially attractive metal for green oxidation catalysts, as it is the least toxic of the transition elements. Moreover, nature usually chooses iron for use in the active site of enzymes that catalyze oxidations with molecular oxygen. Metal-monooxo complexes in the IV and IV-ligand radical cation oxidation states are common in the monooxygenase enzymes where the electronic structures favor considerable reactivity. Monooxygenase processes usually proceed via the formation of peroxide intermediates. But the goal of producing peroxide-activating catalysts with iron makes for a thorny design problem. Iron readily catalyzes Fenton chemistry. Thus, the ligand system employed to achieve an iron-oxo reaction manifold with peroxide must also be resistant to oxidative degradation. If the metal-oxo complex (or any metal-centered reactive intermediate for that matter) oxidatively degrades its own ligand system, not only will the useful lifetime be shortened, but also the iron will be released from its special electronic environment, and Fenton chemistry will ensue. This latter problem can often be ameliorated by adding to catalytic systems chelating ligands that are potent scavengers of free iron, whereby chelation renders the iron inactive towards oxidation processes. But the central catalyst design problem is that the ligand system employed must be resistant to oxidative degradation for a reasonable period if a useful catalyst is going to be achieved.

Twenty years ago, we began the following iterative design process, shown in Figure 1, which was aimed at achieving oxidation-resistant ligand systems that would be useful for making peroxide activating catalysts. Since 1980, we have been studying how to master the ligand decomposition problem (*13, 14*).

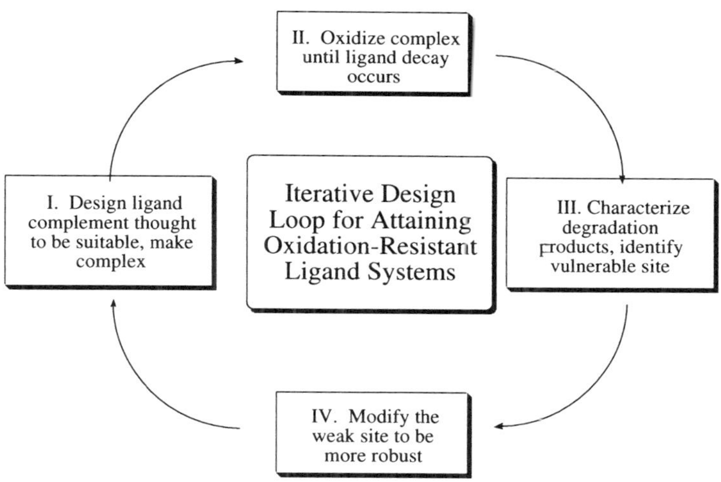

Figure 1. Design loop followed for obtaining oxidation-resistant ligands.

The iterative design strategy proceeds as follows. Step 1: One designs and produces a strongly electron-donating polydentate ligand system thought to be both oxidatively robust and suitable for avoiding Fenton chemistry; a metal is coordinated in a low valent state. Step 2: The complex is chemically or electrochemically oxidized by one or more electrons in an inert medium until the oxidized complex decomposes. Step 3: The degraded system is carefully studied;

complete mass balance is sought to assist in understanding the fate of the ligand and the metal. This study identifies the sensitive group on the ligand where the oxidative degradation process started. Step 4: The vulnerable group is replaced with a substitute thought to be less sensitive, and the cycle is repeated.

This approach has been successful. We have been able to produce catalytic activators of hydrogen peroxide that have viable lifetimes for diverse fields of use, including pulp bleaching (*1, 13–20*). The new catalysts are called TAML (for TetraAmidoMacrocyclicLigand) activators, shown in Figure 2, and they are trademarked in the United States and Europe to Carnegie Mellon University. Ten U.S. patents have been awarded, and internationalization is proceeding with much of this intellectual property.

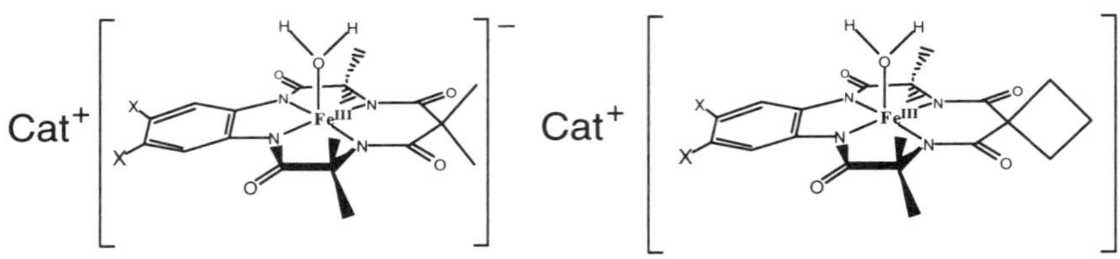

1a; X = Cl
1b; X = H
1c; X = OCH_3

Cat$^+$ = Li^+, $[Me_4N]^+$, $[Et_4N]^+$, $[PPh_4]^+$

2a; X = Cl
2b; X = H

Figure 2. A series of TAML activators.

In our catalyst development program, we have uncovered a deep understanding of how to protect ligands from oxidative decomposition. This understanding is summarized in a series of rules that we are continuing to update (*14*). TAML activators are water-soluble and easy to use and work over a broad pH range. They lead to highly selective chemistry in commercial processes such as pulp bleaching, where selectivity is vital. They are straightforward to synthesize in scalable processes, and they are amenable to modification for capturing novel selectivities. TAML activators are effective in minute concentrations (especially nM to μM) and are capable of more than 10,000 turnovers per hour for certain processes. They work well below the temperatures commonly encountered in chlorine-based counterpart processes, thereby promising large energy savings. The lifetime control we have achieved has given us catalysts with effective lifetimes from seconds to minutes to hours. This suggests the concept of "dial-a-lifetime" catalysts, whereby the catalyst can be chosen such that it will perform the desired task and then be decomposed into inactive material within a period of hours after the catalytic procedure is completed; we believe this is an important green design idea. TAML activators can be comprised exclusively of the common elements of life such that there can be no elemental toxicity. We sometimes use chlorine-substituted activators—the chlorines facilitate catalyst isolation and are not vital components of the reactivity. We anticipate that demand for TAML activators will range across an array of large oxidation-centered fields, leading to economies of scale in production and marketing costs.

The bottom line from TAML activator development is that the principal chemical problem inhibiting access to broadly useful homogeneous catalytic activators of hydrogen peroxide in water has been solved. As with many other fields of use, the P&P industry can look forward to a family of highly effective catalysts for the numerous applications in which hydrogen peroxide is useful. In the remainder of this chapter, we will highlight promising directions for future application of TAML-activated hydrogen peroxide in the P&P industry.

Seven Promises of Oxidation Catalysis for the P&P Industry

In the near future, catalytic peroxide activation has the potential to bring the following advantages to the P&P industry:

1. Increase the selectivity for lignin over cellulose oxidation to give a bleaching process that is more selective than D bleaching.
2. Reduce energy consumption, bleaching time, and operating costs.
3. Reduce or eliminate color, thereby increasing the potential for water recycling while helping to meet environmental and safety regulations.
4. Reduce or eliminate persistent pollutants by not producing them or by catalyzing their destruction, thereby increasing the desirability of water recycling while helping to meet environmental and safety regulations.
5. Improve the efficiency of peroxide usage.
6. Provide special catalysts for unsolved problems.
7. Reduce the capital expenditures required for meeting regulations such as the Cluster Rule (designed to protect human health and the environment by reducing toxic releases to the air and water from U.S. pulp and paper mills).

These promises are more than hypothetical. Some of them present goals that can be commercialized now or in the very near future via TAML activator technology. Others will require further research developments before commercialization can be achieved.

PROMISE 1: Catalysis will lead to the development of TCF processes that have higher selectivity for lignin over cellulose oxidation than all known processes.

TAML activator technology is well on the way to achieving this promise. Laboratory experiments reveal that TAML-activated peroxide is capable of giving the most selective TCF process for pulp bleaching when compared with incumbent TCF processes. Moreover, the selectivity is competitive with D bleaching down to a kappa number (one kappa unit = 0.15% lignin on the pulp by oven dried weight) of approximately 10 (Figure 3).

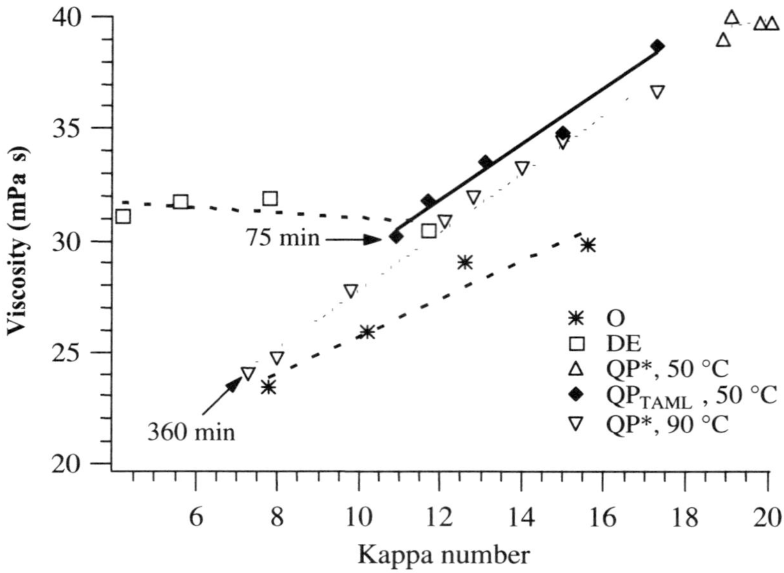

*Figure 3. Selectivity for bleaching of kraft pulp with H_2O_2 (4%) at 50 °C by **2a** (TAML). Q indicates EDTA pretreatment, and an * indicates treatment in the presence of DTMPA. In all cases **2a** (Figure 1, **2a**) was added in 0.1 mg quantities, and the data were collected 15 min after each addition for a total reaction time of 75 min. DE is ClO_2 treatment followed by alkaline extraction.*

Of course, the most forbidding problem of the selectivity at low kappa numbers has yet to be solved; TAML-activated peroxide is still not more selective than D bleaching at low kappa numbers. But with catalyst lifetimes under reasonable control, it is now feasible to develop what we like to call "hunter catalysts," catalysts that have a special affinity for the lignin on the pulp. Such catalysts should improve the selectivity and efficiency of peroxide use, speed up the delignification even further, and lead to a requirement for less catalyst. Thus, our current research efforts are focused partly on achieving effective hunter catalysts.

PROMISE 2: Reduce energy consumption, bleaching time, and operating costs.

Much energy can be saved in TAML bleaching processes versus other processes. The TAML-activated peroxide process occurs rapidly at 50 °C (Figure 3). It is by far the most rapid high selectivity pulp bleaching technology. By comparison, to reach comparable kappa numbers, the unactivated peroxide process requires 3 to 4 hours to achieve comparable delignification. As another indication of the point, activated versus unactivated treatment at 90 °C is shown in Figure 4. At this temperature, the selectivities of the catalyzed and uncatalyzed processes are identical, but the activated process is much faster.

PROMISE 3: Reduce or eliminate color, increase the potential for water recycling, and help meet regulations.

Chlorine dioxide bleaching can be considered to detach lignin from cellulosics in two forms: small fragments, low molecular mass material, and large fragments, high molecular mass material. The effluent then undergoes biological treatment where microorganisms consume the lignin. In general, low molecular mass material is digested, but high molecular mass material is too big to pass through the cell membrane. Color resides on the high molecular mass material, and this makes for a significant pollution problem, because enormous quantities of darkly colored effluent are generated. TAML-activated peroxide provides a technology for dealing with color that is unprecedented in its effectiveness. For example, bleaching of color in a D_C filtrate (D_0 is the first treatment of the pulp with chlorine dioxide) proceeds to 62.5% in 60 min under the conditions shown—almost all the bleaching occurs in the first 5 min (Figure 5). Bleaching of color in an E_{OP} filtrate (E_{OP} is an alkaline washing phase fortified with oxygen and hydrogen peroxide) proceeds to greater than 50% in 30 min under some conditions shown (Figure 6). PAPRO New Zealand has extensively studied the treatment of pulp and paper effluent streams. The technology exhibits significant potential for the removal of color and AOX. The color removal system is robust to fluctuations in operating conditions, needs very low levels of activator, and operates under conditions that should be easy to reproduce in mill applications. This new process is unprecedented in its performance and ease of execution.

PROMISE 4: Reduce or eliminate persistent pollutants by not producing them or by catalyzing their destruction.

TAML-activated peroxide is TCF such that persistent chlorinated pollutants would not be expected to be a part of the effluent profile. Because the activated

PROMISE 5: Improve efficiency of peroxide usage.

The strategy for improving selectivity presented in Promise 1 is also an appropriate strategy to employ to improve the efficiency of peroxide use. By congregating at the molecular level in near proximity to the lignin on the pulp, hunter catalysts should improve the efficiency of peroxide use.

PROMISE 6: Provide special catalysts for unsolved problems.

TAML catalysts are amenable to a furthering of the iterative design process that produced them in the first place. New TAML catalysts will have the potential to orchestrate much more complex reactivity than the prototypes. TAML catalyst lifetime and reactivity can be controlled and the variability in these properties should be amenable to considerable expansion. We are designing TAMLs, called hunter catalysts, to target chemically specific environments in complex heterogeneous substrates such as wood pulp and even to guide reactivity according to prescribed chemical affinity in homogeneous solution. On the issue of lifetime control, we note that there is no virtue in producing an oxidation catalyst that is persistent in the environment. Such a species might eventually find a biological microenvironment in which it can cause mischief (*2*). Because we have a deep understanding of how the activators decompose in use (*13*), we foresee the eventuality of being broadly able to choose an activator with the right lifetime for the oxidation task required; this is the idea inherent in the term "dial-a-lifetime catalyst". Once the task is completed, the activator would conveniently destroy itself in the presence of excess peroxide.

PROMISE 7: Reduce the capital expenditures required for meeting regulations such as the Cluster Rule.

New environmentally friendly technologies that require low capital cost while being effective and inexpensive to operate clearly have the potential to reduce the burden on industry of greening the technology base. TAML-based peroxide industrial processes have desirable features. For example, technologies based on TAML systems will be able to be run much more rapidly, at much lower temperatures, in existing reactors, or in smaller less expensive ones than what the P&P industry is used to dealing with throughout its many processes.

Acknowledgment

We express here our special thanks to Mr. Jonathan Spatz, our Technology Transfer leader, for his many insightful discussions and tireless enthusiasm for TAML technology.

References

1. Collins, T. J.; Horwitz, C. P. In *Proceedings of the 1999 TAPPI Pulping Conference*; TAPPI Press: Orlando, FL, 1999; Vol. 2; pp 703–710.
2. Collins, T. J. *Science* **2001,** *291,* 48–49; *Science* **2001,** *291,* 1899.
3. *Endocrine Disruptors: Effects on Male and Female Reproductive Systems;* Naz, R. K., Ed.; CRC Press: Boca Raton, FL, 1999.
4. *Dioxins and Health*; Schecter, A., Ed.; Plenum Press: New York, 1994.
5. NIEHS; U.S. Department of Health and Human Services Health Service National Toxicology Program, 2001.
6. Thornton, J. *Pandora's Poison: Chlorine, Health, and a New Environmental Strategy;* MIT Press: Cambridge, MA, 2000.
7. Report of the Committee to Review the Health Effects in Vietnam Veterans of Exposure to Herbicides. "Veterans and Agent Orange, Update 1996," Division of Health Promotion and Disease Prevention, Institute of Medicine, National Academy Press: Washington, DC, 1996.
8. Luthe, C. E.; Voss, R. H.; Berry, R. M. *Chemosphere* **1995,** *30,* 661–670.
9. Collins, T. J.; Gordon-Wylie, S. W. *J. Am. Chem. Soc.* **1989,** *111,* 4511–13.
10. Collins, T. J.; Powell, R. D.; Slebodnick, C.; Uffelman, E. S. *J. Am. Chem. Soc.* **1990,** *112,* 899–901.
11. Miller, C. G.; Gordon-Wylie, S. W.; Strazisar, S. A.; Peraino, D. K.; Clark, G. R.; Weintraub, S. T.; Collins, T. J. *J. Am. Chem. Soc.* **1998,** *120,* 11540–11541.
12. Workman, J. M.; Powell, R. D.; Procyk, A.; Collins, T. J.; Bocian, D. F. *Inorg. Chem.* **1992,** *31,* 1548–1550.
13. Bartos, M. J.; Gordon-Wylie, S. W.; Fox, B. G.; Wright, L. J.; Weintraub, S. T.; Kauffmann, K. E.; Münck, E.; Kostka, K. L.; Uffelman, E. S.; Rickard, C. E. F.; Noon, K. R.; Collins, T. J. *Coord. Chem. Rev.* **1998,** *174,* 361–390.
14. Collins, T. J. *Acc. Chem. Res.* **1994,** *27,* 279–285.
15. Collins, T. J.; Gordon-Wylie, S. W.; Bartos, M. J.; Horwitz, C. P.; Woomer, C. G.; Williams, S. A.; Patterson, R. E.; Vuocolo, L. D.; Paterno, S. A.; Strazisar, S. A.; Peraino, D. K.; Dudash, C. A. The Design of Green Oxidants. In *Green Chemistry: Frontiers in Benign Chemical Syntheses and*

Processes; Anastas, P. T., Williamson, T. C., Eds.; Oxford University Press: Oxford, 1998; pp 46–71.

16. Collins, T. J.; Fattaleh, N. L.; Vuocolo, L. D.; Horwitz, C. P.; Hall, J. A.; Wright, L. J.; Suckling, I. D.; Allison, R. W.; Fullerton, T. J. In *TAPPI Pulping Conference;* TAPPI Press: Montreal, 1998; Vol. 3; pp 1291–1300.
17. Collins, T. J.; Hall, J. A.; Vuocolo, L. D.; Fattaleh, N. L.; Suckling, I.; Horwitz, C. P.; Gordon-Wylie, S. W.; Allison, R. W.; Fullerton, T. J.; Wright, L. J. In *Green Chemistry: Challenging Perspectives;* Anastas, P., Tundo, P., Eds.; Oxford University Press: Oxford, 2000; pp 79–105.
18. Hall, J. A.; Vuocolo, L. D.; Suckling, I. D.; Horwitz, C. P.; Allison, R. W.; Wright, L. J.; Collins, T. J. In *Proceedings of the 53rd APPITA Annual Conference*: Rotorua, New Zealand, 1999.
19. Hall, J. A.; Vuocolo, L. D.; Suckling, I. D.; Horwitz, C. P.; Allison, R. W.; Wright, L. J.; Collins, T. J. In *10th International Symposium on Wood and Pulping Chemistry*: Yokohama, Japan, 1999.
20. Horwitz, C. P.; Fooksman, D. R.; Vuocolo, L. D.; Gordon-Wylie, S. W.; Cox, N. J.; Collins, T. J. *J. Am. Chem. Soc.* **1998,** *120,* 4867–4868.

Chapter 5

Spinosad: A Green Natural Product for Insect Control

Gary D. Thompson and Tom C. Sparks

Dow AgroSciences, 9330 Zionsville Road, Indianapolis, IN 46268

Spinosad is manufactured using a new and unique marcolide chemistry produced by fermentation of a naturally occurring soil microorganism. Spinosad's discovery is the direct result of long-standing natural product research programs at Dow Chemical, Eli Lilly and Company, and Dow AgroSciences, all with the goal of discovering greener solutions for agriculture that incorporate high levels of efficacy coupled with benign environmental effects. However, an extraordinary development and manufacturing effort was required to meet the volume and cost hurdles associated with agricultural products. The first product containing spinosad received expedited review by the U.S. Environmental Protection Agency and was granted registration as a "reduced risk" insect control product for cotton in early 1997. Production agriculture is finding that spinosad is a highly effective tool for producing more food and fiber on fewer acres while also protecting the environment. The use of spinosad establishes a new standard for low environmental and human risk and offers a new tool for integrated pest management (IPM) and insect resistance management. Spinosad does not leach, bioaccumulate, volatilize, or persist in the environment. In hundreds of field trials conducted over several years, spinosad left 70 to 90% of beneficial insects unharmed. The insect selectivity, coupled with a novel, rapid mode-of-action on many problem insects, allows a "scout and treat only as needed" approach that results in a powerful IPM tool. Although natural products are not necessarily greener than conventional pesticides and their development can be just as long and costly as traditional synthetic organic insect control agents, spinosad is an explicit example of how natural product discovery based programs can provide greener solutions for agricultural needs.

Introduction

The discovery, characterization and development of spinosad required over twelve years, but the vision of an insect control product that was safe, effective, and based on natural products spanned forty years and two companies. Dow Chemical and Eli Lilly and Company merged their plant science divisions in 1989 to form DowElanco. DowElanco evolved into Dow AgroSciences when Dow Chemical assumed full ownership in 1997. The soil actinomycete *Saccharopolyspora spinosa* and its metabolite spinosad were discovered at Eli Lilly and Company prior to the merger, but there was a parallel natural products effort at Dow. The discovery resulted from a dedicated program with clear product concepts that set hurdle rates for selectivity, cost and customer needs for efficacy on insect targets. Interestingly, the soil sample that produced *S. spinosa* was collected by one of the program's scientists vacationing in the Caribbean, not from professional collections, which is the origin of most samples.

From Discovery to Market

The development of spinosad required a tremendous effort and commitment to overcome a host of technical, perceptive, and developmental issues. Hundreds of employees and scores of teams were involved over a twelve-year time period. As shown in Figure 1, these teams at Dow included Discovery Research, where entomologists designed the novel assays that allowed the discovery and characterization of the initial activity, and natural products chemists who, in a close working relationship with the entomologists, worked to identify the molecule and its complex stereochemical structure. Natural Products Research provided the screening samples, isolation and identification efforts, and initial production. Strain development was an enormous long-term effort, because, not uncommonly, initial titers were very low. Technology

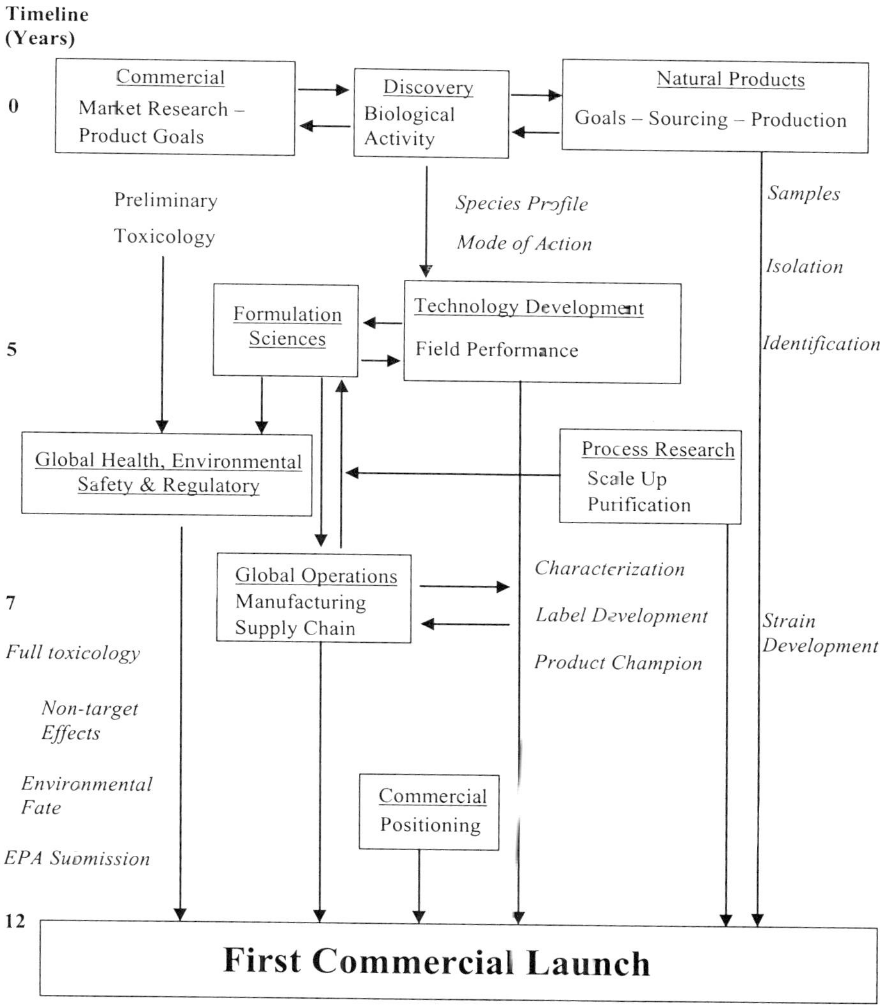

Figure 1. An overview of the organizational complexities and teamwork required to discover and develop spinosad.

Development defined the field performance and use patterns and transferred knowledge to the commercial divisions; they were also the primary product champions and were responsible for justifying continued investment. Formulation Science and Technology developed aqueous-based formulations that maintained the environmental benefits and optimized performance. Process Research did not have to develop chemical reactions, because these take place within the microorganism, but this group had much to do with the manufacturing scale up and the extraction process. The Global Operations division is a collection of smaller departments that detail the important tasks around manufacturing, packaging, supply chain, and quality assurance that are required to get the product to market. Global Health, Environmental Safety and Regulatory coordinated the efforts in toxicology, environmental fate, risk modeling, and regulatory issues. Commercial developed the business and marketing plans, transferred knowledge to the end user, and built awareness and value for green properties, but this group was also involved early on with product goal inputs.

Developing crop protection products is one of the most regulated and complicated product development endeavors. Developing spinosad was even more challenging due to the uniqueness of the chemistry and complexities in manufacturing, and development would have not been possible without a true team effort focused on a common vision. As many of the older insecticides are unable to meet increasing safety standards or to remain economically viable, it is essential that we continue to develop and preserve as many crop protection tools as possible, because high production agriculture is extremely important in maintaining the quality of our environment by reducing land requirements for food and fiber production (*1*).

The Compound

Saccharoployspora spinosa produces a family of novel, secondary metabolites known as the spinosyns. Spinosad (Figure 2) is a mixture of two of the most abundant and lepidopteran-active members of this family, spinosyns A and D (*2*). Structurally, these compounds are macrolides and contain a unique tetracyclic ring system to which two different sugars (forosamine and tri-*O*-methylrhamnose) are attached. A unique mode of action coupled with a high degree of activity on targeted pests and low toxicity to non-target organisms (including many beneficial arthropods) make spinosad an excellent new tool for management of insect pests.

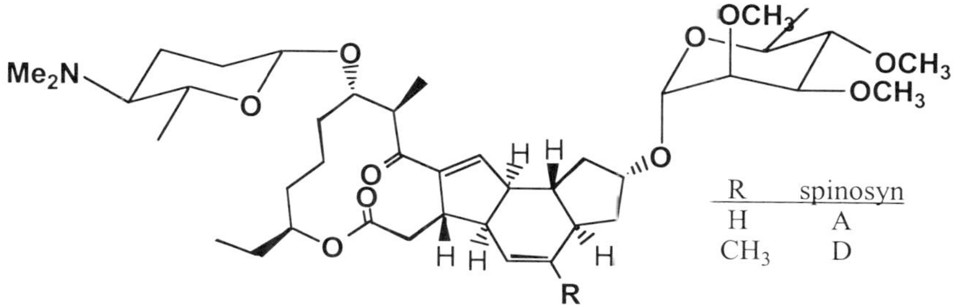

Molecular Weight: spinosyn A = 731.98, spinosyn D = 746.00
Empirical Formula: spinosyn A = $C_{41}H_{65}NO_{10}$, spinosyn D = $C_{42}H_{67}N0_{10}$

Figure 2. Spinosad structure.

Physical Properties and Environmental Fate

Spinosad is a secondary metabolite from the aerobic fermentation of *S. spinosa* on nutrient media. Following fermentation, spinosad is extracted and processed to form a highly concentrated conventional aqueous suspension for ease of use and distribution. Spinosad is a light gray to white crystalline solid with an earthy odor similar to slightly stale water. It has a pH of 7.74 and a shelf life of three years as formulated material, and it is stable and nonreactive. With vapor pressures around 10^{-10} mm Hg, spinosad is considered nonvolatile as well. The degradation of spinosad in the environment occurs through a combination of routes, primarily photodegradation and microbial degradation, ultimately to CO_2, H_2O, and nitrogen oxides. The half-life of spinosad degraded by soil photolysis is 9 to 10 days. The half-life is less than one day for aqueous photolysis, and leaf surface photolysis results in a half-life of 1.6 to 16 days. The half-life of spinosad degraded by aerobic soil metabolism in the absence of light is 9 to 17 days. Hydrolysis does not contribute significantly to degradation, as spinosad is relatively stable in water at a pH of 5 to 7 and has a half-life of at least 200 days at a pH of 9. The leaching potential of spinosad is very low due to a moderate K_d (5 to 323), low to moderate water solubility, and short residual in the environment. Thus, it does not pose a threat to groundwater (*3–5*).

Nontarget Toxicology

Spinosad is relatively low in toxicity to mammals and birds and is only slightly to moderately toxic to aquatic organisms (Table I). In addition, chronic toxicology tests in mammals have shown that spinosad is not carcinogenic, teratogenic, mutagenic, or neurotoxic. Spinosad exhibits wide margins of safety to many beneficial insects and related organisms (*6*). Spinosad has relatively low activity against predaceous beetles, sucking insects, lacewings and mites (Table II). One of the unique characteristics of spinosad is its high level of efficacy against target pest insects coupled with large margins of selectivity for beneficial insect predators, adding an important biological control component to integrated pest management (IPM) programs. Although the topical acute activity of spinosad against honeybees (<1 μg per bee) places spinosad in EPA's highly toxic to bees category, once residues have dried completely, toxicity to foraging bees is considered negligible (*7*). Spinosad couples a high level of activity against target pest insects such as the tobacco budworm (*Heliothis virescens*), a lepidopteran, with the low mammalian activity resulting in an insect versus mammal selectivity ratio that is one of the highest yet developed (Table III) (*8*).

Natural is Not Necessarily Greener

Greener chemistry specifically implies that compounds have either benign, minimal, or greatly reduced environmental or non-target effects, compared to current products. Many people believe that all natural or "organic" products are inherently "green"; however, this is clearly not the case, as demonstrated by some notable exceptions (Table IV). Greener chemistry also implies that the particular molecule possesses insect efficacy at least equivalent to the current products that have inherently more risk. Many other natural or organic insecticides have not been green solutions, because reduced performance or efficacy has resulted in lower yields requiring more planting area or more frequent treatments and environmental injury than non-organic solutions. Typical results of spinosad versus a natural and a synthetic product are illustrated in Figure 3. Spinosad is a natural, organic product that also provides a greener solution.

Physiological Properties and Resistance Management

Spinosad demonstrates rapid contact and ingestion activity in insects, which is unusual for a biological product. The mode of action of spinosad is characterized by excitation of the insect nervous system, leading to involuntary

Table I. Acute Mammalian, Aquatic, and Avian Toxicity of Spinosad

Species	*Test*	*Result*	*EPA Category*
Mammalian			
Rat	Acute oral LD_{50}	3738 / >5000 mg/kg	Caution (IV)
Mouse	Acute oral LD_{50}	>5000 mg/kg	Caution (IV)
Rabbit	Acute dermal LD_{50}	>5000 mg/kg	Caution (IV)
Rat	Acute inhalation LC_{50}	>5 mg/kg	Caution (IV)
Rabbit	Eye irritation	Slight, clearing in 2 days	Caution (IV)
Rabbit	Skin irritation	No irritation	Caution (IV)
Guinea pig	Dermal sensitization	No sensitization	Not applicable
Aquatic			
Daphnia	48 hr acute LC_{50}	92.7 mg/L	Slightly toxic
Grass shrimp	96 hr acute LC_{50}	>9.8 mg/L	Slightly toxic
Carp	96 hr acute LC_{50}	5.0 mg/L	Moderately toxic
Bluegill	96 hr acute LC_{50}	5.9 mg/L	Moderately toxic
Sheepshead minnow	96 hr acute LC_{50}	7.9 mg/L	Moderately toxic
Rainbow trout	96 hr acute LC_{50}	30.0 mg/L	Slightly toxic
Avian			
Bobwhite quail	Acute oral LD_{50}	>2000 mg/kg	Practically non-toxic
Mallard duck	Acute oral LD_{50}	>2000 mg/kg	Practically non-toxic
Bobwhite quail	5 day dietary LC_{50}	>5000 mg/kg	Practically non-toxic
Mallard duck	5 day dietary LC_{50}	>5000 mg/kg	Practically non-toxic

Table II. Toxicity of Spinosad and Cypermethrin to Selected Beneficial Species

Beneficial Species	*Spinosad LC_{50}*	*Cypermethrin LC_{50}*
Honeybee, *Apis mellifera*	11.5 ppm	1.2 ppm
Whitefly parasitoid, *Encarsia formosa*	29.1 ppm	1.9 ppm
Minute pirate bug, *Orius insidiosus*	200 ppm	0.2 ppm
Lady beetle, *Hippodamia convergens*	>200 ppm	0.2 ppm
Lacewing, *Chrysopa rufilabris*	>200 ppm	<0.2 ppm
Predaceous mite, *Phytoseiulus persimilis*	>200 ppm	<0.2 ppm

Table III. Vertebrate Selectivity Ratio for Historical and Many Typical Products for Cotton Pest Control

Compound	*Heliothis virescens* $LD_{50\ (\mu g/g)}$	*Rat Oral* $LD_{50\ (mg/kg)}$	*Vertebrate Selectivity Ratio*
DDT	52–152	87	0.57–1.67
Cyclodienes			
Endrin	46.7	3	0.06
Endosulfan	73.3	18	0.25
Organo-Phosphates			
Methyl Parathion	11.6–65.0	9	0.14–0.77
EPN	16.7–33.0	14	0.42–0.84
Azinphosmethyl	29.3–33.3	5	0.17
Chlorpyrifos	79.5	135	1.70
Monocrotophos	29.7	8	0.27
Acephate	74.3	866	11.70
Profenofos	11.0	400	36.40
Sulprofos	24.0	107	4.46
Carbamates			
Carbaryl	136–232	307	1.32–2.25
Methomyl	4.33–26.7	17	0.64–3.93
Pyrethroids			
Permethrin	1.33–2.79	>4000	>1434–3008
Fenvalerate	0.870–1.89	451	239–1139
Cypermethrin	0.241–1.61	247	153–1025
l-cyhalothrin	0.929	56	60.3
Esfenvalerate	0.429	75	174
Tralomethrin	0.251	1070	4263
Avermectins			
Abamectin	1.16	10.6–11.3	9.1–9.74
Emamectin benzoate	0.10	70	700
Pyrroles			
Chlorfenapyr	4.5	223–459	49.5–100.7
Oxadiazines			
Indoxacarb	0.93	>5000	>5376
Spinosyns			
Spinosyn A	1.28–1.44	3783–5000	2627–3906

NOTE: Vertebrate Selectivity Ratio (VSR) is Acute Insect (*Heliothis virescens*, topical, 48 to 72 hr, µg/g) / Mammalian (rat oral, acute, mg/kg).

SOURCE: Adapted in part from (*9–11*) and from unpublished data.

Table IV. Natural Products with Some Less Than Green Properties

Compound/ Chemical Group	*Source*	*Not So Green Properties*
Rotenone	Plant extract	Rat oral LD_{50} 132–1500 mg/kg, highly toxic to fish and pigs
Physostigmine	Plant extract	Mouse oral LD_{50} 4.5 mg/kg
Nicotine	Plant extract	Rat oral LD_{50} 50–60 mg/kg, high mammalian oral and dermal toxicity
Blasiticidin	Fermentation	Rat oral LD_{50} 56 mg/kg, rather toxic to mammals but rapidly excreted
Abamectin	Fermentation	Rat oral LD_{50} 10 mg/kg, high oral toxicity (technical material) to mammals but low dermal toxicity, very low use rates and formulation result in acceptable safety
Terpenes/Alkaloids	Plant extracts	Secondary defense compounds of plants, certain ones considered highly toxic
Alfalatoxins	Grain	Highly toxic to humans, contaminate grain
Tetrodotoxin	Puffer fish	Highly toxic to humans, mouse IP LD_{50} 10 μg/kg
Palytoxin	Palythoa coral	Extremely toxic to humans, mouse IP LD_{50} 50–100 ng/kg
Botulatoxins	Meat	Highly toxic to humans, food poisoning

muscle contractions, prostration with tremors, and paralysis. Spinosad's mode of action is unique in that it involves the alteration of nicotinic and γ-aminobutyric acid (GABA) receptor functions by a mechanism that is clearly novel among known insect control products. Thus, spinosad does not appear to be either a nicotinic or a "GABAergic" as they are commonly understood. While spinosad does appear to alter the function of both nicotinic and GABA systems, spinosad unmistakably acts via a different mechanism than imidacloprid and other nicotinic receptor-based insecticides. Likewise, spinosad's GABA-related actions are distinct from those of cyclodienes, the fiproles, or abamectin. No other class of products affects the insect nervous system with the same mode of action as spinosad. Bioactivation of spinosad is not required for activity. Equally important, no cross-resistance to spinosad has been demonstrated (*12*).

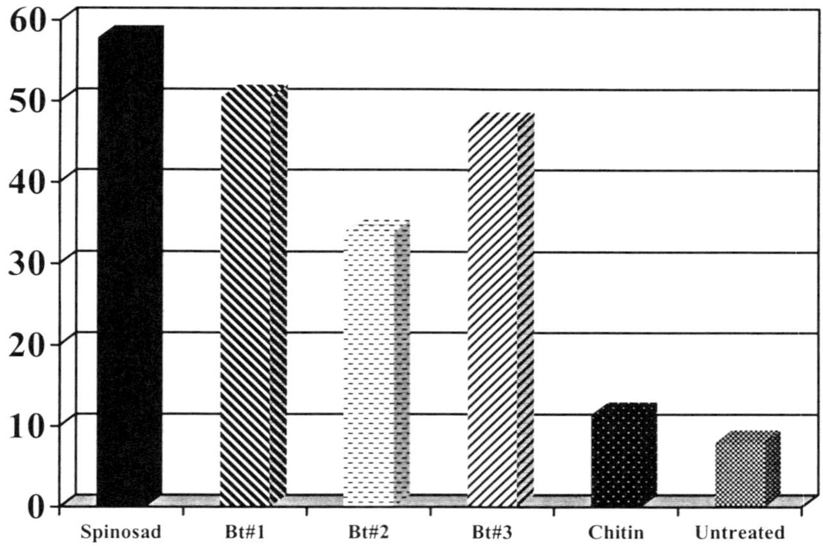

Figure 3. Crop yields of cabbage (lb per 12 heads) in small-plot university tests following applications of spinosad (SpinTor) and other natural insecticides. Adapted from P. Stansly, University of Florida, Arthropod Management Tests 1998, Entomological Society of America; see (21).

Spectrum Of Activity and Labeling Efforts

Spinosad has been tested extensively on a global basis since 1990 (*13–23*). In general, spinosad provides effective control of pests in the insect orders Lepidoptera, Diptera, and Thysanoptera. It is also effective on some species of Coleoptera and Orthoptera that consume large amounts of foliage. Spinosad is generally not effective for control of most sucking insects and mites, but some use patterns are being investigated. Spinosad is currently labeled in the United States on 153 individual crops, including the brassica vegetable group (broccoli, Chinese broccoli, Brussels sprouts, cabbage, Chinese cabbage—bok choy and napa, cauliflower, cavalo, collards, kale, kohlrabi, mizuna, mustard greens, mustard spinach, Chinese mustard cabbage—gai choy, and rape greens), the fruiting vegetable group (eggplant, ground cherry, pepino, pepper, tomatillo, and tomato), the leafy vegetable group (including head and leaf lettuce, celery, arugula, chervil, edible chrysanthemum, corn salad, cress, dandelion, dock, endive, fennel, parsley, garden purslane, radicchio, rhubarb, spinach, and Swiss chard), apples, almonds, citrus, potatoes, sweet corn, the cucurbits crop group, the legume vegetable crop group, grapes, stonefruit, and numerous minor crops.

Conclusions

Spinosad is a new natural product that not only possesses reduced risk attributes, but also delivers functional green solutions for insect control. The Dow AgroSciences team has invested in all aspects of the development, which has resulted in spinosad being rapidly adopted by the agricultural community. It was the product of a targeted insecticide discovery program that overcame significant hurdles with the dedication of individuals and teams that were united by a common vision to provide improved greener solutions for agriculture.

Acknowledgments

Thanks are due to our many colleagues at Eli Lilly and Company and Dow AgroSciences who helped develop spinosad from a research dream into meaningful products. Special thanks are also due N. Orr and G. Watson for discussions on spinosad mode of action.

References

1. Avery, D. T. *Saving the Planet with Pesticides and Plastics;* Hudson Institute: Indianapolis, IN, 1995.
2. Kirst, H. A.; Michel, K. H.; Mynderse J. S.; Chio, E. H.; Yao, R. C.; Nakatsukasa, W. M.; Boeck, L.D.; Occlowitz, J.; Paschel, J.W.; Deeter, J. B.; Thompson, G. D. Discovery, Isolation, and Structure Elucidation of a Family of Structurally Unique, Fermentation-Derived Tetracyclic Macrolides. In *Synthesis and Chemistry of Agrochemicals III;* Baker, D. R., Fenyes, J. G., Steffens, J. J., Eds.; American Chemical Society: Washington, DC, 1992; pp 214–225.
3. Saunders D. G.; Bret, B. L. Fate of Spinosad in the Environment. *Down to Earth* **1997,** *52*, 21–28.
4. DowElanco. *Spinosad Technical Guide;* Indianapolis, IN, 1996; p 25.

5. Copping, L. G., Ed. *The BioPesticide Manual;* The British Crop Protection Council: Farnham, U.K., 1998; pp 38–40.
6. Schoonover, J. R.; Larson, L. L. Laboratory Activity of Spinosad on Non-Target Beneficial Arthropods. *Arthropod Management Tests* **1995,** 357.
7. Mayer, D. F.; Lunden, J. D. Effects of Success on Honey Bee Foraging and Mortality When Applied to Blooming Apples. *Res. Reps. 72nd Ann. Western Orchard Pest & Disease Manage. Conf.* **1998,** 73–74.
8. Hollingworth, R. M. The Biochemical and Physiological Basis of Selective Toxicity. In *Insecticide Biochemistry and Physiology;* Wilkinson, C. F., Ed.; Plenum Press: New York, 1976; pp 431–506.
9. Sparks, T. C. Toxicology of Insecticides and Acaricides. In *Cotton Insects and Mites; Characterization and Management;* King, E. G.; Phillips, J. R., Coleman, R. J., Eds.; The Cotton Foundation: Memphis, TN, 1996; pp 283–322.
10. Sparks, T. C.; Thompson, G. D.; Kirst, K. A.; Hertlein, M. B.; Mynderse. J. S.; Turner, J. R.; Worden, T. V. Fermentation-Derived Insect Control Agents—the Spinosyns. In *Methods in Biotechnology, Vol. 5: Biopesticides: Use and Delivery.* Hall, F. R., Menn, J. J., Eds.; Humana Press Inc.: Totowa, NJ, 1999, pp 171–188.
11. Harder, H. H.; Riley, S. L.; McCann, S. F.; Irving, S. N. DPX-MP062: A Novel Broad-Spectrum, Environmentally Soft, Insect Control Compound. In *Proc. Brighton Crop Protection Conf.* The British Crop Protection Council: Farnham, U.K., 1996; pp 449–454.
12. Salgado, V. L. The Mode of Action of Spinosad and Other Insect Control Products. *Down to Earth* **1997,** *52*, 35–44.
13. Carson, W. G.; Trumble, J. T. Effect of Insecticides on Celery Insects. *Arthropod Manag. Tests* **1997,** *22*, 117.
14. Fouche, C.; Canevari, M.; Cutter, D. Evaluation of Insecticides for Control of Leafminers on Lima Beans. *Arthropod Manag. Tests* **1997,** *23*, 74–75.
15. Kerns, D. L. Control of Lepidopterous Larvae and Leafminers in Lettuce. *Arthropod Manag. Tests* **1995,** *21*, 117–118.
16. Linduska, J. J.; Ross, M.; Baumann, D.; Parr, A. Foliar Sprays to Control Ear-Invading Insects on Sweet Corn. *Arthropod Manag. Tests* **1997,** *23*, 95–96.
17. McLeod, P. Evaluation of Insecticides for Control of Corn Earworm on Snap Bean. *Arthropod Manag. Tests* **1998,** *23*, 75.
18. Palumbo, J. C. Evaluation of Selective Insecticides for Control of Lepidopterous Larvae in Lettuce. *Arthropod Manag.* **1997,** *22*, 136.
19. Riley, D. G. Evaluation of Insecticide Treatments on Cabbage. *Arthropod Manag. Tests* **1998,** *23*, 82.
20. Schuster, D. J. Management of Insects on Fresh Market Tomatoes. *Arthropod Manag. Tests* **1997,** *22*, 182.

21. Stansly, P. A.; Connor, J. M. Impact of Insecticides Alone and in Rotation on Tomato Pinworm, Leafminer and Beneficial Arthropods in Staked Tomatoes. *Arthropod Manag. Tests* **1998,** *23*, 162–165.
22. Walgenbach, J. F.; Palmer, C. R. Control of Lepidopterous Insects on Cabbage. *Arthropod Manag. Tests* **1997,** *22*, 113.
23. Webb, S. E. Control of Pickleworm on Squash with Selective Insecticides. *Arthropod Manag. Tests* **1998,** *23*, 142–143.

Chapter 6

Toward Benign Synthesis via Catalytic Oxidations Using Dioxygen or Nitrous Oxide

Klaus H. Theopold[1], Olivia M. Reinaud[2], Sebastien Blanchard[3], Somying Leelasubcharoen[4], Alexandra Hess[5], and Sunita Thyagarajan[1]

[1]Department of Chemistry and Biochemistry, Center for Catalytic Science and Technology, University of Delaware, Newark, DE 19716
[2]Current address: Laboratoire de Chimie et Biochimie, Université René, Descartes (Paris V), 75270 Paris, France
[3]Current address: Laboratoire de Chimie Inorganique, I.C.M.O., Université de Paris-Sud, Orsay, France
[4]Current address: Department of Chemistry, Khon Kaen University, Khon Kaen 40002, Thailand
[5]Current address: Gebrüder Haake GmbH, 76227 Karlsruhe, Germany

Oxidation reactions comprise a large class of important chemical transformations. The use of O_2 or N_2O as terminal oxidants is attractive from an environmental viewpoint, because of the lack of harmful byproducts and the utilization of a waste stream and greenhouse gas in the latter case. The principal problem with this approach is the kinetic inertness of these molecules. Our research is concerned with activating dioxygen or nitrous oxide by binding to transition metals, and ultimately creating homogeneous catalysts for selective oxidations.

Introduction

O_2 Activation

In elucidating the fundamental mechanistic patterns of air oxidations of organic molecules, two complementary approaches can be taken. The first of these is to "activate" the dioxygen—that is, transforming it into a reactive species—to the point where it will react in an intermolecular fashion with unactivated organic substrates. A possible sequence of transformations toward that end is depicted in equation 1. The crucial point appears to be that the O–O bond of O_2 must be broken; another is that the metal designated by M should be a late transition metal in a relatively high formal oxidation state (*1*).

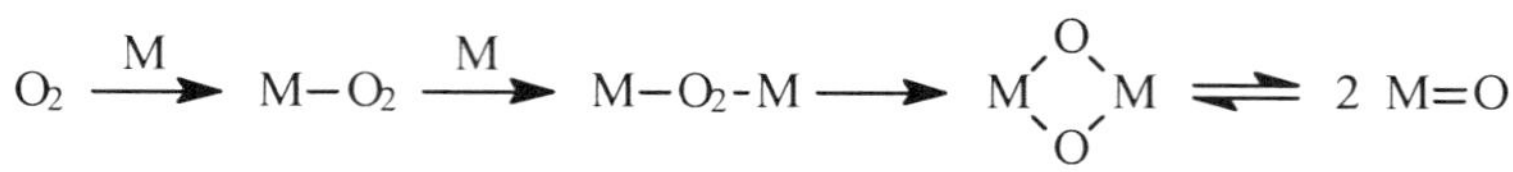
(1)

An alternative approach is the activation of the organic substrate, for example via formation of an organometallic compound, which could subsequently react with O_2. While the mechanisms of such reactions have not been fully elucidated, equation 2 depicts a possible series of steps leading to oxygenative functionalization of a metal–carbon bond.

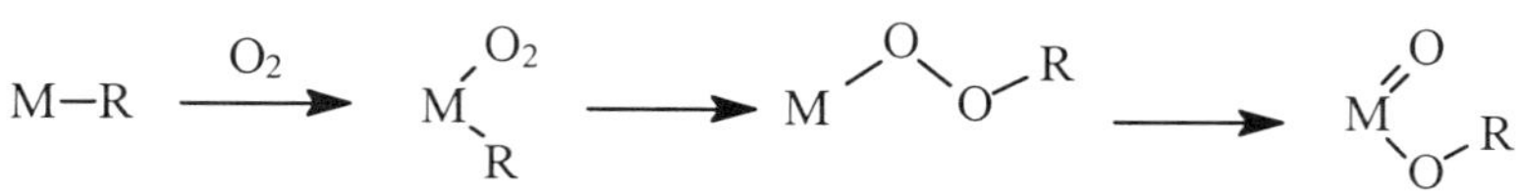
(2)

To lay the foundation for the design of catalytic cycles for the air oxidation of organic molecules, we are exploring both kinds of processes. The following sections summarize some results of our relevant studies using cobalt and chromium complexes.

N_2O Activation

The activation of nitrous oxide by transition metals proceeds in all probability through an unstable N_2O complex, which extrudes N_2 to generate a potentially reactive metal oxo complex (equation 3) (*2*). An example supporting this assertion is described in the last part of the next section.

(3)

Results

O_2 Activation with Cobalt Complexes

We are using sterically encumbered tris(pyrazolyl)borate cobalt complexes in an attempt to create an artificial oxidation catalyst along the lines of equation 1. The first step was the preparation of the dioxygen complex $Tp^{t\text{-}Bu,Me}Co(O_2)$, the first example of a side-on bound superoxide complex (*3*). $Tp^{t\text{-}Bu,Me}Co(O_2)$, while thermally stable by itself, was found to react with a second equivalent of metal—in the form of $Tp^{t\text{-}Bu,Me}Co(N_2)$—to yield $Tp^{t\text{-}Bu,Me}Co\text{-}OH$, the ostensible product of O–O cleavage and hydrogen atom abstraction. As a test of this mechanistic hypothesis, the analogous $Tp^{i\text{-}Pr,Me}Co$-system was prepared. $Tp^{i\text{-}Pr,Me}Co(O_2)$ decomposes at ambient temperature, yielding the binuclear hydroxide, in which hydrogen has been abstracted from the isopropyl-substituents of the ligand. However, the reaction proceeds through a spectroscopically observable intermediate; based on several lines of evidence, the formula $Tp^{i\text{-}Pr,Me}Co(\mu\text{-}O)_2CoTp^{i\text{-}Pr,Me}$ has been assigned to this intermediate.

The decomposition of this complex involves hydrogen abstraction from the ligand; it exhibits an unusually large kinetic isotopic effect, which has been rationalized by invoking quantum mechanical tunneling (*4*). The surprising difference in the thermal stabilities of $Tp^{t\text{-}Bu,Me}Co(O_2)$ and $Tp^{i\text{-}Pr,Me}Co(O_2)$ has been traced to the accessibility of a dimer of the latter, that is, $[Tp^{i\text{-}Pr,Me}Co(O_2)]_2$, which can lose O_2 to form the dinuclear peroxo-complex (*5*). The whole sequence of transformations is shown in Scheme 1.

We have used density functional calculations to determine the structures of reaction intermediates and to estimate bond energies. Thus, our structural assignment of $Tp^{i\text{-}Pr,Me}Co(\mu\text{-}O)_2CoTp^{i\text{-}Pr,Me}$ was based on calculations, although the crystal structure of a similar molecule has recently been reported and confirms the assignment (*6*). The dissociation of dinuclear $Tp^{i\text{-}Pr,Me}Co(\mu\text{-}O)_2CoTp^{i\text{-}Pr,Me}$ into two mononuclear cobalt oxo complexes is energetically very unfavorable.

Abstraction Resistant Ligands

One major problem that must still be solved in transforming the chemistry outlined above into a functioning oxidation catalyst is the oxidative degradation of the Tp-ligands. To direct the reactivity of the bis(μ-oxo)species toward external substrates, we must "harden" the ligands against the prevalent hydrogen atom abstraction from the substituents in the 3-position of the pyrazoles. It seemed to us that a promising approach to the solution of this problem might be the use of perfluoro alkyl substituents. The great bond strength of the C–F bond (105 to 108 kcal/mol) should render such groups impervious to attack by the oxo ligands and should hopefully redirect their reactivity elsewhere. There is some previous work on perfluoroalkyl substituted Tp-ligands, and we have begun a study of their use in this context (*7*).

The trifluoromethyl substituted $Tp^{CF3,CF3}$-ligand unfortunately proved too unhindered to enforce the necessary tetrahedral geometry of cobalt. While we have prepared $Tp^{CF3,CF3}CoI$, it showed a tendency to gradually disproportionate to octahedral $(Tp^{CF3,CF3})_2Co$. Figure 1 shows the molecular structure of $Tp^{C2F5}CoI$, the next candidate in this series, featuring the slightly larger perfluoroethyl groups.

We have not yet had any success in reducing this complex to the corresponding dinitrogen complex, and it remains to be seen whether this approach will ultimately be fruitful. One possible problem may be the strong electron-withdrawing nature of the perfluoroalkyl groups, which stabilizes the Tp-anion to the point of being a reasonable leaving group. However, this is only the first attempt at addressing the larger problem of protecting our ligand system against intramolecular attack.

O_2 Activation with Chromium Complexes

Another system that accomplishes the activation of molecular dioxygen is the chromium complex $[Cp^*Cr(\mu\text{-}Cl)(Cl)]_2$. Upon exposure to O_2 gas, the latter rapidly and quantitatively yields $Cp^*Cr(O)Cl_2$, a high-valent organometallic chromium(V) oxo complex. This chemistry parallels an earlier observation of analogous reactivity of $[Cp^*Cr(\mu\text{-}Br)(Br)]_2$ (*8*). The $Cp^*Cr(O)Cl_2$ complex has been fully characterized, including a crystal structure determination, the result of which is shown in Figure 2.

Scheme 1

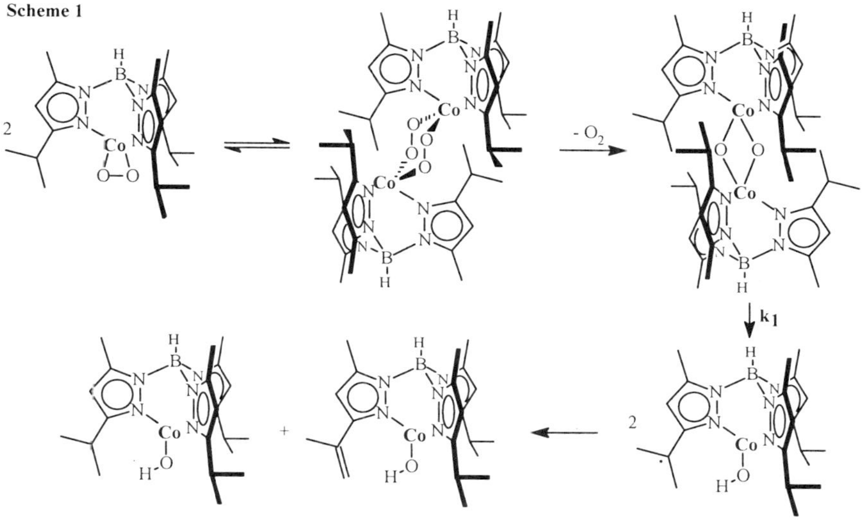

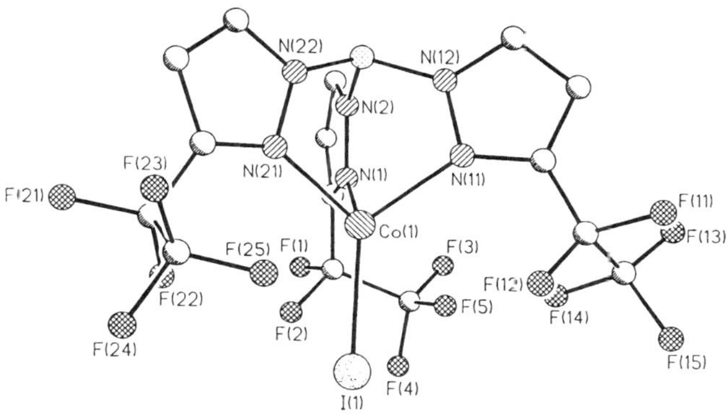

Figure 1. The molecular structure of TpCo^{C2F5}CoI.

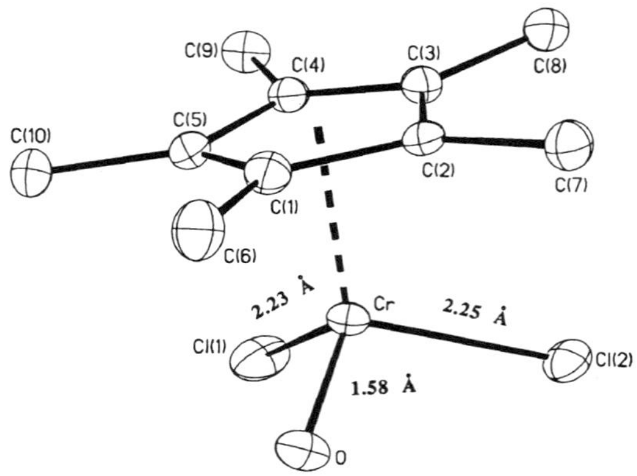

*Figure 2. The molecular structure of Cp*Cr(O)Cl₂.*

In light of its easy accessibility and the intense interest in catalytic processes utilizing O_2 as terminal oxidant, we have investigated the reactions of Cp*Cr(O)Cl_2 with oxidizable substrates. Cp*Cr(O)Cl_2 is capable of transferring oxygen atoms to easily oxidizable molecules (e.g., phosphines, arsines), and it can activate weak X–H (X=C, N) bonds and dehydrogenate selected organic molecules (H_2O being the byproduct). Some of these reactions can be run catalytically in the presence of excess O_2, for example, the oxidation of triphenylphosphine and the dehydrogenation of 1,4-cyclohexadiene. However, this Cr^V oxo complex fails to transfer oxygen to less electron rich substrates (e.g., olefins, amines, CO), and it does not dehydrogenate or oxygenate common hydrocarbons.

The epoxidation of olefins by metal oxo complexes is a well-precedented reaction, but efficient catalysts for the 'direct oxidation' (i.e., the code for 'epoxidation using O_2 as a reagent') of alkyl-substituted olefins (e.g., propene) are not known. The major problem of this transformation is the susceptibility of allylic C–H bonds to radical attack, providing competing reaction/decomposition pathways. Cp*Cr(O)Cl_2 does not react with olefins, such as norbornene, ethylene or cyclohexene. However, the expected products of these reactions, that is, epoxide complexes of the type Cp*Cr(OC_2R_4)Cl_2 (R = H, alkyl) can be formed by reaction of [Cp*CrCl(μ-Cl)]$_2$ with the appropriate epoxide. Thus it was not clear whether the oxygen atom transfer reaction fails for thermodynamic or kinetic reasons. This question has been addressed by DFT calculations, however. The calculations revealed that the oxygen atom transfer from the chromium complex to ethylene—to form the ethylene oxide adduct Cp*Cr(OC_2H_4)Cl_2—is thermodynamically favorable (ΔG = -12.2 kcal) but faces a substantial activation barrier. Further research will be aimed at modifying the chromium complex and thereby decreasing the activation energy for the epoxidation.

Guided by the computational studies, we have, for example, prepared the analogous $CpCr(O)Cl_2$. The computations predict both a more exothermic oxygen atom transfer to ethylene as well as a lower barrier for this reaction. Indeed, our experiments are consistent with these predictions, that is, $CpCr(O)Cl_2$ reacts with certain olefins to produce epoxides and/or their decomposition products. However, the activation of dioxygen by $[CpCrCl(\mu\text{-}Cl)]_2$ is extremely slow, probably due to the insolubility of both molecules. We are now preparing molecules that will deal with this problem, in the hope of developing a catalyst for the O_2 driven epoxidation of olefins.

Reaction of Organometallics with O_2

Representative of the approach outlined in equation 2 is our recent study of the reaction of $Tp^{t\text{-Bu,Me}}Cr\text{-}Ph$ with O_2 (*9*). The product of this reaction is paramagnetic $Tp^{t\text{-Bu,Me}}Cr^{IV}(O)OPh$ (see Figure 3), that is, the product of a formal insertion of oxygen into the metal-carbon bond. While C_2 insertion into metal alkyls are precedented (*10*), they are often thought to proceed by radical chain mechanisms. Mechanistic investigations showed, however, that this reaction is intramolecular (no crossover!) and proceeds via an IR-spectroscopically observable O_2 complex, that is, $Tp^{t\text{-Bu,Me}}Cr(O_2)Ph$. This reaction appears to be a rare example of a metal-mediated direct insertion of O_2 into a metal-carbon bond. The insertion of oxygen into metal-carbon bonds is of interest as a fundamental step in metal-catalyzed oxidations of hydrocarbons; thus the transformation noted above is a possible step in a chromium-mediated direct oxidation of benzene to phenol, a reaction of great commercial interest.

Activation of N_2O with Cobalt

We reasoned that the failure of our cobalt-based O_2-activation scheme to produce a mononuclear cobalt complex with a terminal oxo ligand might be circumvented by the use of a powerful single oxygen atom donor, such as N_2O. Hence we investigated the reaction of $Tp^{t\text{-Bu,Me}}Co(N_2)$ with N_2O. There was an immediate reaction, although we could not obtain any evidence for an intact N_2O complex, even by in situ IR spectroscopy at low temperatures. Upon completion of the reaction, we found two inorganic products formed in a 4:1 ratio. The major product was the hydroxide complex $Tp^{t\text{-Bu,Me}}Co\text{–}OH$, and the other product resulted from a functionalization of one of the *tert*-butyl C–H bonds, as shown in Scheme 2.

The detailed mechanism of this reaction is obviously somewhat complex, involving a hydrogen atom transfer in a bimolecular reaction. However, it is apparent that the interaction of N_2O with the cobalt atom generates a species reactive enough to break primary C–H bonds and form a new C–O bond. While the functionalization of the ligand, rather than an external substrate, remains a

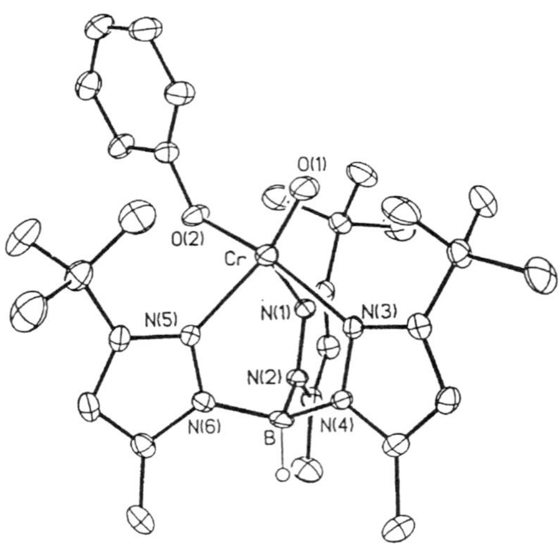

Figure 3. The molecular structure of $Tp^{t\text{-}Bu,Me}Cr(O)OPh$*; Cr-O(1), 1.58Å; Cr-O(2), 1.84Å.*

vexing problem. The concurrent development of abstraction resistant Tp-ligands may eventually solve this problem.

Scheme 2

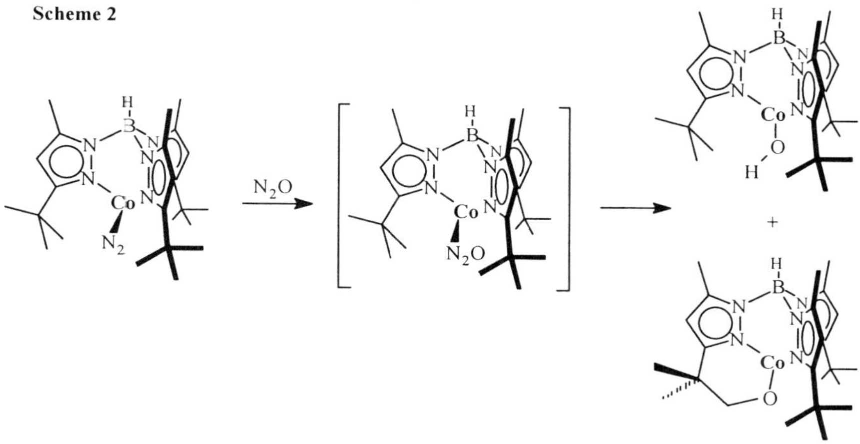

Conclusions

We have shown that various metal complexes can be used to activate dioxygen, that is, to coordinate it and split it into reactive metal oxo moieties. The oxygenation of preformed metal-carbon bonds can be accomplished by reacting organometallics with O_2, and we have demonstrated that the mechanism of O_2 insertion does not have to be a radical chain process.

Oxidation catalysis using O_2 as a reagent remains an attractive option from an environmental standpoint. In general, catalytic processes are preferable over stoichiometric oxidations, because the latter often generate large amounts of toxix wastes (e. g., high-valent chromium compounds). O_2 is readily available and does not require synthesis; the omission of a chemical process to prepare an oxidant represents a savings in energy and avoids the generation of potentially harmful byproducts. When carried out in the most atom economic way, that is, assuming the utilization of both oxygen atoms of O_2, there remain no byproducts to dispose of. Finally, a natural product of O_2 oxidations is the environmentally benign substance water (H_2O).

On the subject of N_2O as a terminal oxidant, we have demonstrated that metal catalysts can readily transform this molecule into an extremely active oxidant. One of the remaining challenges of this chemistry is the protection of the organic ligands against self-degradation. Our studies represent a contribution to the fundamental understanding of homogeneous oxidation catalysis, and we will continue to search for promising catalysts for the utilization of O_2 and N_2O as practical oxidants.

Acknowledgments

These studies were carried out with the support of grants from the U.S. Department of Energy (ER14273) and the National Science Foundation (CHE-9876426). O. M. Reinaud received a stipend from the Centre National de la Recherche Scientifique (CNRS), and S. Leelasubcharoen was the recipient of a fellowship from the Royal Thai government.

References

1. a) Tolman, W. B. *Acc. Chem. Res.* **1997**, *30*, 227–237; b) Akita, M.; Moro-Oka, Y. *Catal. Today* **1998,** *44,* 183–188.
2. For an analogous mechanism for the decomposition of alkyl azide complexes see: Proulx, G.; Bergman, R. G. *Organometallics*, **1996,** *15,* 684–692.
3. Egan, J. W. Jr.; Haggerty, B. S.; Rheingold, A. L.; Sendlinger, S. C.; Theopold, K. H. *J. Am. Chem. Soc.* **1990,** *112,* 2445–2446.
4. Reinaud, O. M.; Theopold, K. H. *J. Am. Chem. Soc.* **1994,** *116,* 6979–6980.
5. Reinaud, O. M.; Yap, G. P. A.; Rheingold, A. L.; Theopold, K. H. *Angew. Chem., Int. Ed. Engl.* **1995,** *34,* 2051–2052.
6. Hikichi, S.; Yoshizawa, M.; Sasakura, Y.; Akita, M.; Moro-Oka, Y. *J. Am. Chem. Soc.* **1998,** *120,* 10567–10568.
7. a) Dias, H. V. R.; Kim, H.-J. *Organometallics* **1996**, *15*, 5374–5379; b) Dias, H. V. R.; Jin, W. *Inorg. Chem.* **1996**, *35*, 3687–3694; c) Dias, H. V. R.; Jin, W. *Inorg. Chem.* **1996**, *35*, 267–268; d) Dias, H. V. R.; Kim, H.-J.; Lu, H.-L.; Rajeshwar, K.; de Tacconi, N. R.; Derecskei-Kovacs, A.; Marynick, D. S. *Organometallics* **1996**, *15*, 2994–3003; e) Dias, H. V. R.; Wang, Z.; Jin, W. *Inorg. Chem.* **1997**, *36*, 6205–6215; f) Hu, Z.; Williams, R. D.; Tran, D.; Spiro, T. G.; Gorun, S. M. *J. Am. Chem. Soc.* **2000**, *122*, 3556–3557; g) Gorun S. M.; Hu, Z.; Stibany, R. T.; Carpenter, G. *Inorg. Chim. Acta* **2000**, *297*, 383–388; h) Fekl, U.; Van Eldik, R.; Lovell, S.; Goldberg, K. I. *Organometallics* **2000**, *19*, 3535–3542.
8. Morse, D. B.; Rauchfuss, T. B.; Wilson, S. R. *J. Am. Chem. Soc.* **1988**, *110*, 8234–8235.
9. Hess, A.; Hörz, M. R.; Liable-Sands, L. M.; Lindner, D. C.; Rheingold, A. L.; Theopold, K. H. *Angew. Chem., Int. Ed. Engl.* **1999**, *38*, 166–168.

10. a) Kochi, J. K. *Organometallic Mechanisms and Catalysis*, Academic Press: New York, **1978**, pp 517–522; b) Blackburn, T. P.; Labinger, J. A.; Schwartz, J. *Tet. Lett.* **1975**, 3041–3044; c) Ryan, D. A.; Espenson, J. A. *J. Am. Chem. Soc.* **1979**, *101*, 2488–2489; d) Lubben, T. V.; Wolczanski, P. T. *ibid.* **1987**, *109*, 424–435; e) Van Asselt, A., Trimmer, M. S.; Henling, L. M.; Bercaw, J. E. *ibid.* **1988**, *110*, 8254–8255; f) Parkin, G.; Bercaw, J. E. *ibid.* **1989**, *111*, 391–393.

Chapter 7

An Environmentally Benign Catalytic Polyoxometalate Technology for Transforming Wood Pulp into Paper

Ira A. Weinstock[1], Elena M. G. Barbuzzi[2], Dan M. Sonnen[3], and Craig L. Hill[2]

[1]Forest Service, Forest Products Laboratory, U.S. Department of Agriculture, One Gifford Pinchot Drive, Madison, WI 53705–2398
[2]Department of Chemistry, Emory University, 1515 Pierce Drive, Atlanta, GA 30322
[3]Forest Products Laboratory, University of Wisconsin at Madison, One Gifford Pinchot Drive, Madison, WI 53705–2398

An environmentally benign technology, based on a solution of equilibrated polyoxometalate (POM) salts and oxygen, is being developed to bleach wood pulps for use in the manufacture of paper. Unlike conventional chlorine-based technologies, this biomimetic chemistry uses only oxygen and water and produces no liquid waste. This research constitutes a new general approach to the design of soluble transition metal catalysts for aerobic oxidation in water. The POM-based catalyst is thermodynamically stable under the operating conditions and is the first catalyst designed to act as *its own buffer*.

Introduction

Wood is composed of structural polysaccharides (primarily cellulose) and cross-linked methoxylated phenyl propane polymers (lignin). Cellulose is the most abundant constituent of the cell wall and is the key component of the natural fiber from which paper is made. It imparts strength to trees and papers. Natural and industrial processes alike focus on the removal of lignin, which is the primary encrusting cell wall matter in higher plants, with minimal damage to cellulose (high-quality paper is composed of lignin-free cellulose fibers).

Delignification of wood (*1*) is a remarkably complex process (*2*). At the same time, it is one of the largest-scale chemical transformations in nature and in industry. The most desirable bleaching conditions involve removal of lignin with little or no damage to the cellulose and other polysaccharides. A number of factors make delignification very difficult: the intrinsically complex organic structure of lignin, the presence in the wood cell wall structure of covalent linkages between lignin and polysaccharide, and the complex morphology of the cell wall with its physical and structural integration of lignin- and polysaccharide-rich structural domains on molecular to macroscopic scales. Biological delignification (via fungi) can be highly selective (*3–5*) and is environmentally friendly. Lignin-degrading fungi secrete metalloenzymes, which oxidatively cleave the lignin polymer to water-soluble fragments via single-electron transfer steps. These extracellular enzymes include Cu-containing oxidases (laccases) (*6*) and heme-containing peroxidases (lignin and Mn peroxidases) (*7, 8*). Once the lignin polymer is broken down into water-soluble fragments, additional enzymes convert these fragments to CO_2 and H_2O (*9*).

This chapter discloses the first technology that selectively delignifies pulp fibers in water using only O_2 (*10*) and produces no liquid waste or toxic byproducts (Scheme 1) (*11–15*). This technology involves the use of a multi-component, transition metal oxygen-anion cluster (polyoxometalate, or POM), $Na_6[AlVW_{11}O_{40}]$ (henceforth **1**).

Polyoxometalates

POMs are an important and rapidly growing class of inorganic cluster compounds (*16*). The most common and most investigated class of POMs are the Keggin compounds. The Keggin structure is comprised of MO_6 octahedra linked together by one or two bridging oxygen atoms. The complex, **1**, is formed by substituting one of the d^0 W(VI) ions that makes up the inorganic skeleton of a parent Keggin compound, $Na_5[AlW_{12}O_{40}]$, by a d^0 V(V) ion.

While POMs range in size from 0.9 nm to over 5 nm, the Keggin structure has a diameter of about 1.1 nm, one similar to that of typical phenyl propane units in lignin.

POMs are attractive for use in the bleaching process for several reasons. First, they are easily prepared in a single step in H_2O from inexpensive, minimally toxic, readily accessible precursors such as WO_3 and $NaVO_3$. Second, their physical properties, including redox potentials, acidities, charges, and solubilities, can be controlled by the choice of precursors and synthetic conditions. Third, POMs are resistant to oxidation (only d^0 metal ions are present), yet they can be reversibly reduced by lignin and a host of other organic substrates (*17*). Finally, one or more of the d^0 metal ions (W(VI) ions in the case of **1**) can be replaced by other d^0 or d-electron containing transition-metal ions.

As a result, the appropriately optimized POMs can selectively remove electrons from organic substrates such as lignin and facilitate their transfer to O_2. As such, the POM in our technology performs the same function as does the ensemble of extracellular enzymes in wood-rotting fungi (Scheme 1).

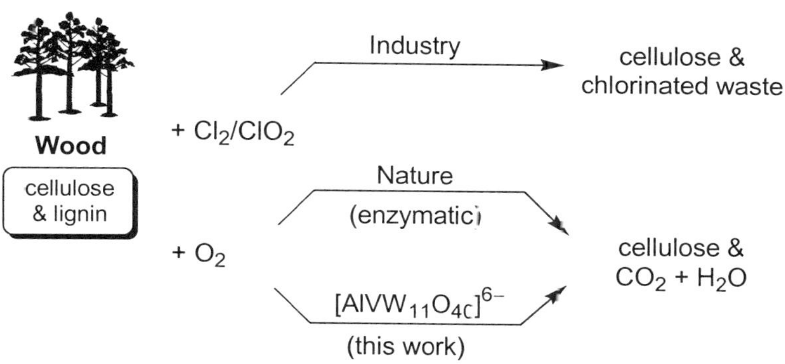

Scheme 1

POM Bleaching

Pulp bleaching by POMs is a two-step technology that is being developed at the Forest Products Laboratory of the U.S. Department of Agriculture and the Department of Chemistry at Emory University. The selective removal and mineralization of lignin by a solution of **1** and O_2 is accomplished in two steps that sum to the selective transfer of electrons from lignin to O_2 (Scheme 2).

In the first step (step A, Scheme 2), a slurry of unbleached pulp fiber is heated anaerobically in an aqueous solution of the fully oxidized catalyst, **1**. Electron transfer from lignin converts **1** to its reduced form, $Na_7[AlVW_{11}O_{40}]$ (**2**). As in nature (degradation by fungal enzymes), the lignin is cleaved by single-electron transfer and subsequent hydrolysis reactions. Prior to step B, the high quality (high viscosity, etc.) delignified fiber is collected. In step B, **2** is reoxidized to **1** by O_2, and the combined reaction of **1**, **2**, and dioxygen results in the complete oxidative mineralization of the dissolved lignin fragments to CO_2 and H_2O (*18*). As in all organic oxidation reactions in H_2O, protons are generated during substrate oxidation (step A). During the reduction of O_2 to water, these protons are consumed such that over a complete cycle (steps A and B), the concentration of H^+ in the system remains unchanged.

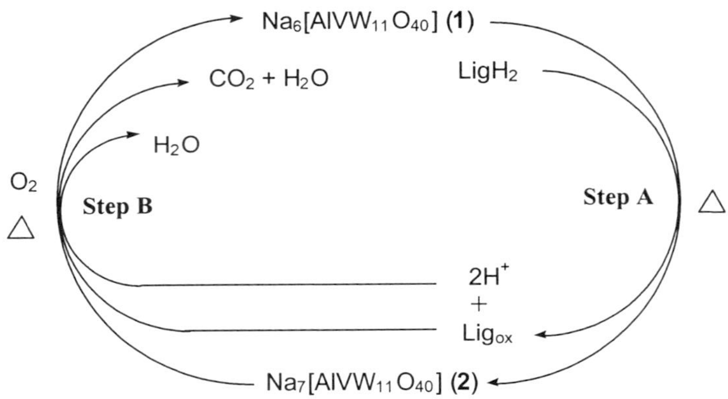

$LigH_2$ = lignin

Scheme 2

However, to avoid acid-catalyzed hydrolysis of the α-D-glucopyranoside linkages in cellulose during delignification, a pH buffer is needed in step A. In previous work we demonstrated that $[SiV^{V}W_{11}O_{40}]^{5-}$ is effective in step A, but it is rapidly degraded by the phosphate buffer needed to maintain the pH of the solution near neutral (*14*). Moreover, the reoxidation of $[SiV^{IV}W_{11}O_{40}]^{6-}$ by dioxygen (step B) is very slow, precluding effective catalytic turnover (steps A + B). We also evaluated the POM $[PV_2Mo_{10}O_{40}]^{5-}$ in early studies because the reduced form was known to be reoxidized quickly by reaction with O_2.

Unfortunately, this POM is stable only at the acidic pH values at which cellulose hydrolysis occurs, and thus the pulp does not survive step A (*14*). The catalyst reported here is an equilibrium mixture of POM clusters in which **1** is the major product. It is capable of sustained activity at near-neutral pH values in both the steps A and B. The ability of POM to maintain such pH values in the absence of an exogenous buffer involves acid-base reactions associated with the self-assembly synthesis of the POM itself. We refer to this unprecedented ability of a redox reagent to maintain the pH of its own redox reactions as 'self-buffering'. The equilibrated POM-based catalyst system in this chapter is not the most efficient system for the bleaching process; other POMs (mixed Mo and W systems) provide better results. We have focused on equilibrated solutions of **1** because they afford three effective complementary spectroscopic handles for detailed and quantitative analysis of the chemistry: ^{27}Al, ^{51}V and ^{183}W NMR (*19*).

This system based on the POM **1** exhibits most of the features of an ideal environmentally benign oxidation catalyst: (1) O_2 is the only oxidant used; (2) water is the only solvent required; (3) the catalyst, **1**, possesses no organic ligands or components and is consequently thermodynamically stable toward oxidative degradation; (4) a single molecule, **1**, facilitates the repeated operation of multiple functions: delignification (bleaching) and catalytic wet oxidation of the dissolved lignin fragments (aerobic mineralization); (5) the catalytically active species, **1**, is in equilibrium with its synthetic precursors under operation conditions (a rare situation in catalysis); should it be degraded during turnover, its optimal form is spontaneously reconstituted; and (6) the catalyst system (**1**, its precursor, and other species present at equilibrium) functions as its own pH buffer (i.e., the system is self-buffering).

Experimental Approach

General

All materials used were reagent grade. The pulp was an unbleached mixed-pine kraft pulp with a kappa number of 32.4 and a viscosity of 31.1 mPa•s. The bleaching and wet oxidation reactions were performed in a 300-mL Parr 4843 Reactor. The UV-vis spectra were acquired using a Hewlett-Packard 8452A spectrophotometer. All pH measurements were made using an Orion model 250A pH meter.

Acquisition of ^{27}Al ^{51}V ^{183}W NMR Spectra

The spectra were obtained using a Varian UNITY 400 NMR at 16.66 MHz. External references were, for ^{27}Al, 0.10 M $AlCl_3•6H_2O$ ($[Al(H_2O)_6]^{3+}$ $\delta = 0$ ppm), for ^{51}V, 10 mM $H_4[PVMo_{11}O_{40}]$ in 0.60 M NaCl ($\delta = -533.6$ ppm), and for ^{183}W, 0.2 M $Na_2WO_4•2H_2O$ (WO_4^{2-}(aq), $\delta = 0$ ppm).

Results and Discussion

It is possible to obtain an isomerically pure sample of **1** by the standard method of POM synthesis: sequential kinetically controlled acid condensation and/or base-hydrolysis steps. Importantly, it is also possible to obtain equilibrium solutions of **1** by preparing the catalyst at the concentration and temperature at which it is designed to operate. The rationale behind this approach is that a catalyst system prepared in this manner should be thermodynamically stable under turnover conditions. Eq. 1 gives the stoichiometry for the equilibrium self-assembly synthesis of **1** from simple metal oxide precursors. This reaction is carried out in a single hydrothermal step in which the synthetic precursors are heated at 200 °C for 12 hours in water.

$$Al(OH)_3 + NaVO_3 + 11\ WO_3 + 5\ NaOH \rightarrow Na_6[AlVW_{11}O_{40}] + 4\ H_2O \qquad (1)$$

Use of the stoichiometry in eq. 1 provides an equilibrated solution that contains **1** as the major product (95% relative to the total vanadium present (0.25M V(V))), along with a small concentration $Na_5[AlW_{12}O_{40}]$ (mixture of α and β isomers, 5% relative to the total aluminum present (0.25 M Al(III))), $Na_5[V_3W_3O_{19}]$ and $Na_4[V_2W_4O_{19}]$ (both 1% relative to the total vanadium present). The pH resulting from the stoichiometry in eq. 1 is 3.9, a pH too low for use in delignification. To increase the pH and to optimize the self-buffering properties, the concentration of each component was varied systematically. The optimized stoichiometry, given in eq. 2, was achieved by increasing the relative ratio of $Al(OH)_3$, WO_3, and NaOH.

$$0.725\ Al(OH)_3 + 0.500\ NaVO_3 + 6.50\ WO_3 + 3.20\ NaOH \rightarrow$$
$$0.475\ Na_6[AlVW_{11}O_{40}]\ (\mathbf{1}) + 0.012\ Na_4[V_2W_4O_{19}] + 0.050\ Na_6[Al_2W_{11}O_{39}] +$$
$$0.050\ Na_7[Al_3W_{10}O_{38}] + 0.025\ Na_6[W_7O_{24}] + 2.69\ H_2O \qquad (2)$$

A key to this chemistry is the aluminum concentration in solution. The aluminum concentration must be sufficient to avoid precipitation of the tungsten species and simultaneously must be adequate to facilitate the formation of species important in the self-buffering process. The amount of tungsten was also optimized to achieve the highest concentration of catalyst **1**. Additional POM species in solution were quantified by ^{27}Al and ^{51}V NMR (see Figures 1 and 2).

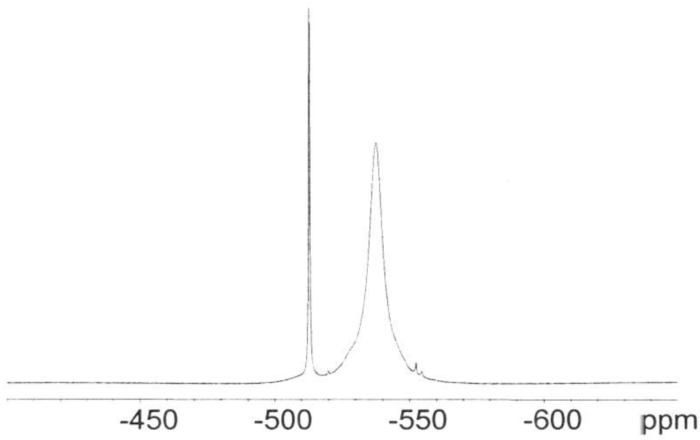

Figure 1. ^{51}V NMR spectrum of 1 prepared by hydrothermal synthesis. The minor peak at -511 ppm is $Na_4[V_2W_4O_{19}]$, and the major peak at -536 ppm is $Na_6[AlVW_{11}O_{40}]$.

When softwood kraft pulp (kappa number of 32.4 and a viscosity of 31.1 mPa•s) is heated as a slurry in an equilibrated 0.5 M solution of **1** at 130 °C for 130 min under anaerobic conditions (argon atmosphere) to eliminate non-selective autoxidation, 78% of the residual lignin in the unbleached pulp fibers is oxidatively depolymerized to water soluble fragments, while about 15% of the **1** in solution is reduced to **2**. The decrease in the lignin component of the pulp fibers was assessed by kappa number, and the damage to cellulose was assessed by viscosity. After minimum optimization, the low kappa number pulps (kappa numbers of 6) obtained possessed final viscosities of 23 mPa•s. These values are comparable to those typically obtained industrially using chlorine compounds and far exceed those possible using O_2 (final viscosities of less than 15 mPa•s).

This new catalyst system is designed for continuous circulation of the POM solution. Before the POM is used in a new bleaching step, however, **2** is reoxidized to **1** by O_2 under conditions that result in complete mineralization of dissolved lignin fragments to CO_2 and H_2O (see Scheme 2). The POM solution used in the delignification experiment described above was reoxidized by heating at 210 °C for three hours under O_2 (2 MPa). As a model for continuous operation, this same POM solution was used to perform seven additional bleaching-mineralization cycles with fresh samples of unbleached pulp used in each bleaching step. The results are summarized in Figures 3 and 4. The CO_2 produced was measured during each of the eight cycles and compared with the amount expected based on the masses of lignin removed and the yields of delignified pulp samples (Figure 3). It is noted that during each bleaching step, the pH of the POM solution remained nearly constant throughout the eight cycles.

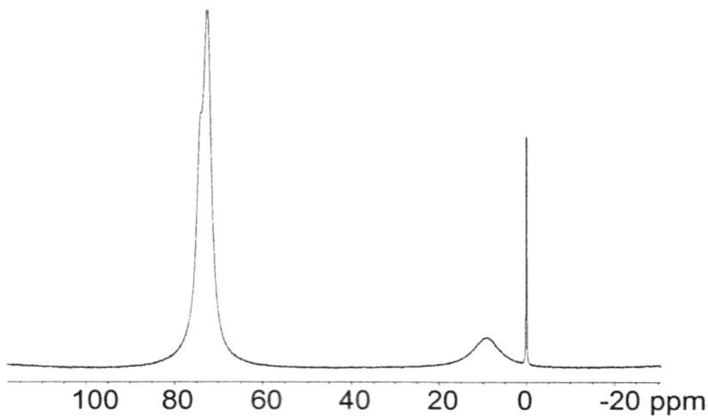

*Figure 2. ^{27}Al NMR spectrum of **1** prepared by hydrothermal synthesis. The peak at 0 ppm is $AlCl_3$ (external standard), the peaks at 8.5 ppm and 73.5 ppm are $Na_6[Al_2W_{11}O_{39}]$, and the one at 73.2 ppm is $Na_6[AlVW_{11}O_{40}]$.*

If left unbuffered, the pH would have dropped to about 1.2 during each bleaching reaction (the theoretical drop of the pH is simply based on the percentage reduction of the POM solution). It was independently determined that each molecule of **1** was reduced by one electron to form **2,** and **2** only, with the simultaneous generation of one equivalent of H^+.

The consumption of H^+ ions introduced during oxidation of lignin by **1** in step A and of OH^- anions formed upon reduction of O_2 by **2** in step B is due to the reversible reaction of metalate species summarized in eq. 3. $Na_4[V_2W_4O_{19}]$, $Na_6[Al_2W_{11}O_{39}]$ and $Na_6[AlVW_{11}O_{40}]$ (**1**) were quantified using ^{51}V and ^{27}Al NMR (Figures 5 and 6).

$$4\ H^+ + Na_4[V_2W_4O_{19}] + Na_6[Al(Al)W_{11}O_{39}] + Na_6[W_7O_{24}] \Leftrightarrow 2\ Na_6[AlVW_{11}O_{40}]\ (\mathbf{1}) + 2\ H_2O + 4Na^+ \quad (3)$$

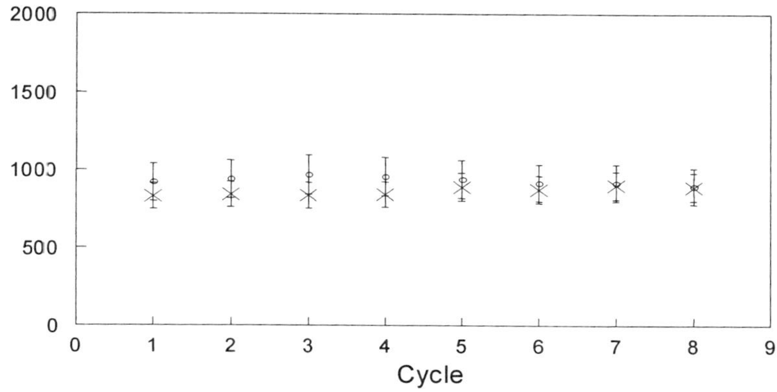

*Figure 3. Quantification of the CO_2 produced by aerobic mineralization of dissolved lignin fragments during reoxidation of **2** to **1** in 8 delignification/reoxidation cycles. The theoretical CO_2 values were calculated for each point from a combination of pulp yields and decreases in lignin content. Experimental $[CO_2]$ values were determined by reaction of head gases with solutions of $Ba(OH)_2$.*

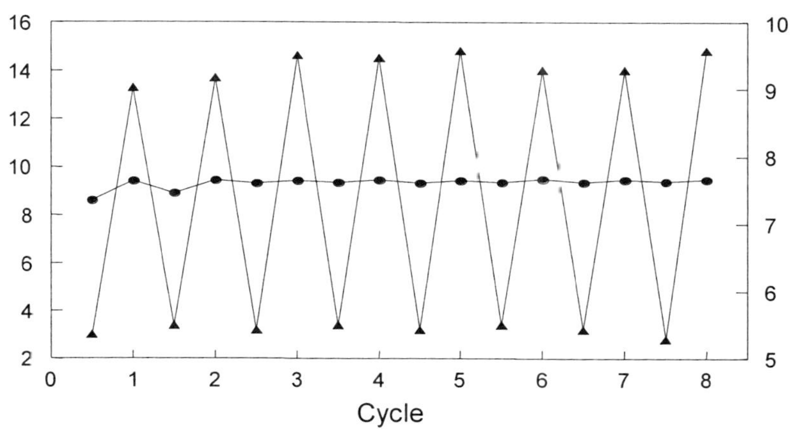

*Figure 4. A single 0.5 M solution of $Na_6[AlVW_{11}O_{40}]$ (**1**) used repeatedly in successive delignification and O_2-reoxidation (8 complete cycles). These two steps (delignification plus reoxidation) constitute one complete cycle shown on the abscissa. Concentrations of $Na_7[AlVW_{11}O_{40}]$ (**2**) and pH were determined after each step from UV-visible spectroscopy ($\varepsilon_{(680nm)}$ = 180 $cm^{-1}M^{-1}$) (left scale percent reduction = ▲; right scale pH = ●).*

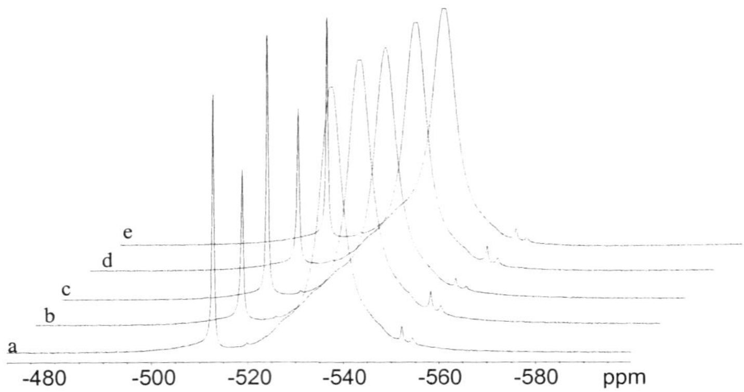

Figure 5. ^{51}V *NMR of (a) a solution of freshly prepared* ***1****, (b) the solution after reduction by lignin in cycle 1, (c) after reoxidation by* O_2 *in cycle 1, (d) after the final delignification step (cycle 8), (e) after final reoxidation (cycle 8).*

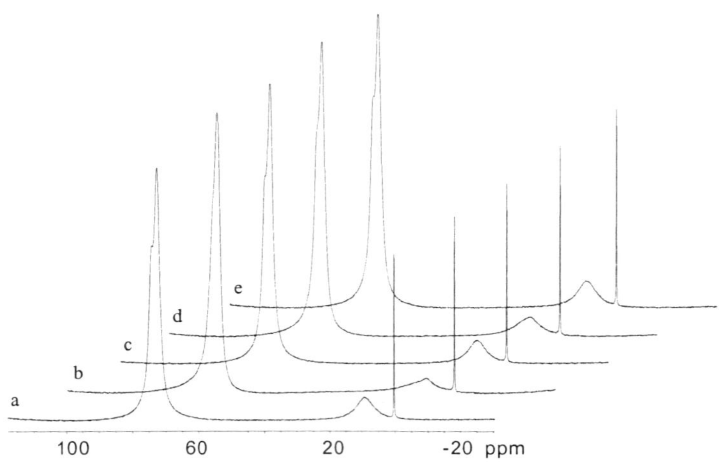

Figure 6. ^{27}Al *NMR of (a) a solution of freshly prepared* ***1****, (b) the solution after reduction by lignin in cycle 1, (c) after reoxidation by* O_2 *in cycle 1, (d) after the final delignification step (cycle 8), (e) after final reoxidation (cycle 8).*

Eq. 3 explains most of the chemistry responsible for the self-buffering. However, complete delineation of this complicated chemistry (quantitation of all POM species present including those in trace amounts) is extraordinarily difficult. The POM speciation responsible for self-buffering is satisfactorily modeled by reduction of a 0.5 M solution of **1** by CO. Use of CO, unlike heterogeneous reactions with pulp, results in a 100% reduction of **1**. Thus study with CO oxidation facilitated the quantification of the POM species responsible for this internal pH management chemistry. In the absence of self-buffering, the pH at 100% of reduction would have been 0.3. During reduction of a 0.5 M solution of **1** by CO, the pH remains near neutral (40% reduction) and then gradually decreases to pH 6 at 50%, 5 at 70% and 4 at 100%. Clearly this POM system has considerable self-buffering ability. The products from reduction of **1** by CO (200 °C and 5 hours) are CO_2 and H_2O, just as when lignin was used as the substrate.

After a careful analysis of the completely reduced and oxidized solution by ^{27}Al, ^{51}V and ^{183}W NMR (see representative experiments in Figures 7 and 8), we were able to better understand and rationalize the self-buffering properties. $Na_4[V_2W_4O_{19}]$ is a limiting reagent in eq. 3. During complete reduction of **1**, it disappears by reaction with $Na_6[Al_2W_{11}O_{39}]$. The remaining $Na_6[Al_2W_{11}O_{39}]$ is consumed by reaction with paratungstate (eq. 4).

$$Na_6[Al_2W_{11}O_{39}] + (^1/_7)Na_6[W_7O_{24}] + (^{34}/_7)H^+ \Leftrightarrow$$
$$Na_5[AlW_{12}O_{40}] + Al^{3+} + (^{13}/_7)Na^+ + (^{17}/_7)H_2O \qquad (4)$$

Eq. 4 by itself accounts for ~50% of the H^+ consumed by the POM system. After complete reduction, the concentration of $Na_7[Al_3W_{10}O_{38}]$ changes slightly. It was possible to rationalize this behavior via eqs. 5 through 8:

$$Na_7[Al_3W_{10}O_{38}] + Na^+ \Leftrightarrow Na_{10}[Al_2W_{10}O_{38}] + Al^{3+} \qquad (5)$$

$$7\ Na_{10}[Al_2W_{10}O_{38}] + Na_6[W_7O_{24}] + 34H^+ \Leftrightarrow$$
$$7\ Na_9[Al_2W_{11}O_{39}] + 7\ Al^{3+} + 15\ Na^+ + 17\ H_2O \qquad (6)$$

$$Na_9[Al_2W_{11}O_{39}] + (^1/_7)Na_6[W_6O_{24}] + (^{34}/_7)H^+ \Leftrightarrow$$
$$Na_5[AlW_{12}O_{40}] + (^{34}/_7)Na^+ + (^{17}/_7)H_2O \qquad (7)$$

$$Al^{3+} + (^{12}/_7)Na_6[W_7O_{24}] + (^{16}/_7)H^+ \Leftrightarrow Na_5[AlW_{12}O_{40}] + (^{37}/_7)Na^+ + (^8/_7)H_2O \qquad (8)$$

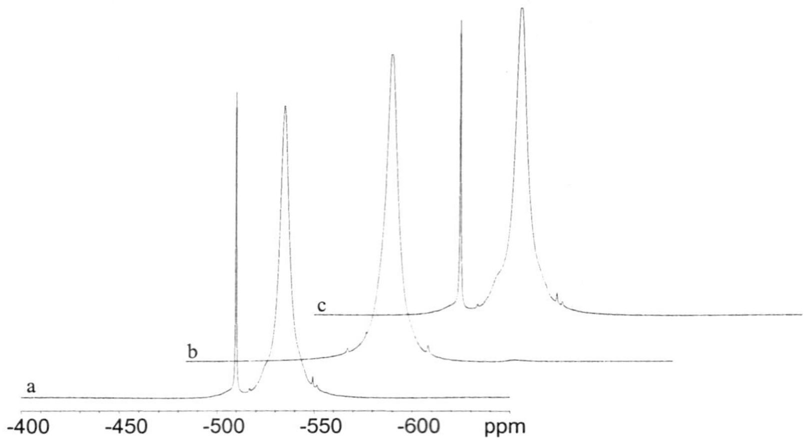

*Figure 7. ^{51}V NMR of (a) a solution of freshly prepared **1**, (b) the solution after complete reduction by CO, and (c) the solution after the complete subsequent reoxidation.*

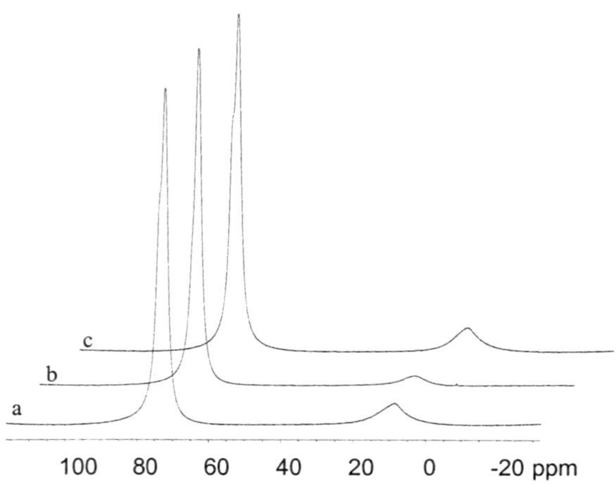

*Figure 8. ^{27}Al NMR of (a) a solution of freshly prepared **1**, (b) the solution after complete reduction by CO, and (c) the solution after the complete subsequent reoxidation.*

Conclusions

An environmentally friendly and economically attractive technology, using POMs, for the conversion of wood pulp into paper has been described. Unlike the current chemistry that uses corrosive chlorine compounds and generates considerable toxic chlorinated organic waste, this chemistry produces no liquid waste or toxic effluent.

The approach used is unique: an equilibrated solution of POMs, simply using oxygen as the oxidant and water as the solvent, is able to facilitate the selective delignification of wood pulp, generating only H_2O and CO_2 as byproducts.

Acknowledgments

This research was supported by the U.S. Department of Agriculture (Forest Products Laboratory), the U.S. Department of Energy (FPL & Emory University) and the National Science Foundation (Emory University). We thank Jennifer Cowan for manuscript preparation.

References

1. Fengel, D.; Wegener, G. *Wood: Chemistry, Ultrastructure, Reactions*; Walter de Gruyter: New York, 1984.
2. Sjostrom, E. *Wood Chemistry: Fundamentals and Applications*; 2nd ed.; Academic Press: San Diego, CA, 1992.
3. Srebotnik, E.; Messner, K. *Appl. Environ. Microbiol.* **1994**, *60*, 1383.
4. Srebotnik, E.; Jensen, J., K. A.; Hammel, K. E. *Proc. Natl. Acad. Sci.* **1994**, *91*, 12794.
5. Kersten, P. J.; Kalyanaraman, B.; Hammel, K. E.; Reinhammer, B.; Kirk, K. T. *Biochem. J.* **1990**, *268*, 475.
6. Kersten, P. J. *Proc. Natl. Acad. Sci. USA* **1990**, *87*, 2936.
7. Tien, M.; Kirk, T. K. *Science* **1983**, *221*, 661.
8. Glenn, J. K.; Gold, M. H. *Arch. Biochem. Biophys.* **1985**, *242*, 329.
9. Higuchi, T. *Biosynthesis and Biodegradation of Wood Components*; Academic Press: San Diego, CA, 1985.
10. Hill, C. L.; Weinstock, I. A. *Nature (London)* **1997**, *388*, 332–333; Hill, C. L. *Nature (London)* **1999**, *401*, 436–437.
11. Weinstock, I. A.; Hill, C. L. U.S. Patent 5,302,248, 1994 and subsequent patents in 1996, 1997, and 1998.

12. Weinstock, I. A.; Atalla, R. H.; Reiner, R. S.; Moen, M. A.; Hammel, K. E.; Houtman, C. J.; Hill, C. L. *New J. Chem.* **1996**, *20*, 269–275.
13. Weinstock, I. A.; Atalla, R. H.; Reiner, R. S.; Moen, M. A.; Hammel, K. E.; Houtman, C. J.; Hill, C. L.; Harrup, M. K. *J. Mol. Catal. A: Chem.* **1997**, *116*, 59–84; Weinstock, I. A.; Barbuzzi, E. M. G.; Wemple, M. W.; Cowan, J. J.; Reiner, R. S.; Sonnen, D. M.; Heintz, R. A.; Bond, J. S.; Hill, C. L. *Nature.* **2001**, in press.
14. Weinstock, I.; Atalla, R.; Reiner, R.; Houtman, C.; Hill, C. *Holzforschung* **1998**, *52*, 304–310.
15. Weinstock, I.; Hammel, K.; Moen, M.; Landucci, L.; Ralph, S.; Sullivan, C.; Reiner, R. *Holzforschung* **1998**, *52*, 311–318.
16. Pope, M. T.; Müller, A. *Angew. Chem. Int. Ed.*, **1991**, *30*, 34; Pope, M. T.; Müller, A. *Polyoxometalates: From Platonic Solids to Anti-retroviral Activity*, Kluwer: Dordrecht, 1993, *Chem. Rev.*, (special issue on polyoxometalates), Hill, C. L., Ed. **1998**, *98*, 1–390.
17. Weinstock, I. A. *Chem. Rev.* **1998**, *98*, 113–170.
18. Sonnen, D. M.; Reiner, R. S.; Atalla, R. H.; Weinstock, I. A. *Ind. Eng. Chem. Res.* **1997**, *36*, 4134–4142.
19. Weinstock, I. A.; Cowan, J. J.; Barbuzzi, E. M. G.; Zeng, H.; Hill, C. L. *J. Am. Chem. Soc.* **1999**, *121*, 4608–4617.

Chapter 8

Enhancement of Herbicide and Insecticide Activity with Thermal Polyaspartate

Ramon Georgis, Robert J. Ross, and Larry P. Koskan

Donlar Biosyntrex Corporation, 6502 South Archer Avenue, Bedford Park, IL 60501

Thermal polyaspartate (TPA) is commercially available as a nutrient absorption enhancer in agricultural applications. Recent investigations have shown that certain variations of TPA enhance the uptake, and therefore the effectiveness, of foliar and soil insecticides, as well as pre-emergence and post-emergence herbicides. The practicality of such benefits was demonstrated successfully in greenhouse and field tests. TPA is compatible with most common insecticides and herbicides and also can be tank-mixed and applied with conventional sprayers.

Introduction

Thermal polyaspartate (TPA) is a biodegradable, water-soluble polypeptide obtained via the thermal polymerization of aspartic acid (*1*). TPA has a variety of uses in water treatment as a dispersant and inhibitor of mineral scale deposition (*2–4*), in oil production as a scale and corrosion inhibitor (*5*), in personal care product manufacture (*6, 7*), in agriculture as a nutrient absorption enhancer (*8, 9*), and in a variety of pharmaceutical applications (*11–13*).

Aspartic acid, when heated to a temperature in excess of 180 °C, undergoes a solid-state condensation polymerization to afford the useful polymeric intermediate known as polysuccinimide. Treatment of polysuccinimide with an aqueous base, such as sodium hydroxide, affords sodium poly-α,β-D,L-aspartate, also known as thermal polyaspartate. Acid catalysts, such as phosphoric acid, have been added to the aspartic acid to afford higher molecular weight polysuccinimide than is obtained in the non-catalyzed polymerization.

Thermal polyaspartate has been shown to enhance yield and nutrient uptake in agricultural crops when applied with fertilizers. Commercial products are currently sold in various market segments. Enhancement of root growth, root-hair length, and root-hair longevity have been reported in many crops by various researchers (*9*). Recent field and laboratory tests have shown that TPA enhances the uptake, and therefore the efficiency, of insecticides and herbicides. The potential to optimize the uptake of both pesticide groups promises better pest control as well as a reduction of chemicals disseminated to the environment, including ground-water chemicals.

Biological Characteristics

Thermal polyaspartate has a high capacity to attract cations due to the negative charges associated with the carboxyl functional groups of TPA molecules. The mixture of TPA and pesticides creates a concentrated environment of charged molecules that apparently leads to faster and more efficient penetration of active pesticides into insects or weeds. TPA used to enhance pesticide activities has the following characteristics:

- Chemical name: copoly [(3-carboxypropionamide) (2-(carboxymethyl) acetamide)], sodium salt.
- Common name: Carpramid; L-aspartic acid, homopolymer, sodium salt.
- Chemical structure:

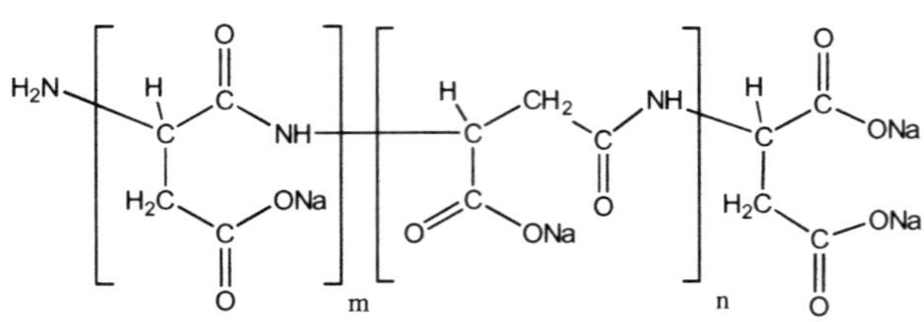

- Color: yellow to amber liquid.
- pH: 7 to 10.
- Specific gravity: 1.25 to 1.40 for a 40% solution in water.
- Flash point: none.

- Freezing point: -10 °C to –15 °C.
- Stability: very soluble in water.
- Molecular weight: 3,000 to 5,000.

Thermal polyaspartate is extremely stable at room temperature. Its molecular weight remains unchanged for at least four years at ambient room temperature and for nine months at 50 °C. Under extreme conditions, TPA exposed to pH 11 to 13 remained stable at 65 °C and 95 °C for 3 and 18 hours, respectively. At room temperature, TPA exposed to pH 5 and pH 1 remained stable for 12 and 24 weeks, respectively.

Ecological and Environmental Fate

Under aerobic-soil environments, TPA is biodegradable. The rate of degradation depends on the physical and the biological conditions of the soil (*14*). In a laboratory study, microbial degradation of TPA at a 100 ppm rate was determined in ten different soil types using ^{14}C-TPA (*15*). Biodegradation after 25 days ranged from 2.5% in pH 5.1 sandy-loam soil to 32% in pH 7.8 sandy-loam soil. Regression analysis indicated that biodegradation was highly and positively correlated to the soil pH regardless of soil type. The activity of complex enzymes from various bacteria species were likely the major factors in TPA degradation in alkaline soil. In another study (*15*), it was shown that complex microbial communities and relatively high C:N ratios optimize mineralization of TPA. Accordingly, flora and nutrient composition represent additional degradation-controlling factors for TPA that would vary in natural-soil environments. In the same study, TPA mobility in seven soil types was highly correlated with absorption; the mobility was less in silty clay and silty-loam soils than in sandy soils.

Various toxicological tests showed that TPA is safe on a wide range of organisms. A summary of the results includes:

- Acute oral LD_{50} (rat) >5,000 mg/kg
- Acute inhalation LC_{50} (rat) >5.1 mg/L
- Eye irritation (rabbit) non irritant
- Skin irritation (rabbit) non irritant
- *Daphnia pulex* (waterfleas) LC_{50} >1700 ppm
- *Pimephales promelas* (fathead minnow) LC_{50} >2000 ppm
- *Skeletonema costatum* (algae) LC_{50} >1000 ppm

Biological Data—Insecticides

Several bioassay techniques and greenhouse tests were performed to evaluate the benefit of TPA in enhancing the efficacy of various insecticides. The addition of TPA to insecticides applied to soil or foliage led to increases in their effectiveness to insects. In most cases, the use of TPA with lower than the label rate of insecticides provided equivalent or better control of the target insect than their label rate (Tables I through IV). Also, in many applications, the addition of TPA to the label rate of certain insecticides enhanced their speed of action and/or effectiveness against certain target insects (Tables II through IV).

Table I. Enhancing Insecticidal Effectiveness against Imported Fire Ant *Solenopsis invicta*.

Treatment	*% Mortality of Imported Fire Ants*	
	0.25 Label Rate	*0.125 Label Rate*
Diazinon	70.8 (1 hr)	53.4 (2 hr)
Diazinon + TPA	91.0 (1 hr)	69.2 (2 hr)
Sevin	---	18.7 (1 hr)
	---	80.7 (2 hr)
Sevin + TPA	---	80.7 (1 hr)
	---	94.0 (2 hr)

NOTE: Treatment is 5 L/ha label rate for Diazinon and Sevin. Mortality was recorded at 1 or 2 hours after contact between the insect and the insecticide.
SOURCE: Laboratory studies *(15)*.

Table II. Enhancing Success (Spinosad) Effectiveness against Tobacco Budworm *Heliothis viescens*.

	% Mortality of Tobacco Budworm Larvae	
Success + TPA	*24 HAT*	*43 HAT*
0.10 Label Rate	13.6 a	47.1 b
0.10 Label Rate + 5 L/ha	21.7 a	86.3 a
0.00 Label Rate + 5 L/ha	1.7 b	5.6 c
Untreated (Check)	2.3 b	9.9 c

NOTE: Means followed by the same letters are not significantly different at the 0.05 level. Label rate is 0.2 kg/ha, foliar application. HAT is hours after treatment.
SOURCE: Laboratory studies (1999).

Table III. Enhancing Success (Spinosad) Effectiveness against Cabbage Looper *Trichoplusiani*.

Success + TPA	*% Mortality of Cabbage Looper Larvae*	
	17 HAT	*41 HAT*
0.10 Label Rate	46.2 c	97.3 a
0.10 Label Rate + 2 L/ha	90.0 a	96.6 a
0.10 Label Rate + 5 L/ha	72.1 b	93.1 a
0.00 Label Rate + 5 L/ha	2.5 d	5.3 b
Untreated (check)	1.9 d	10.2 b

NOTE: Means followed by the same letters are not significantly different at the 0.05 level. Label rate is 0.2 kg/ha, foliar application. HAT is hours after treatment.
SOURCE: Laboratory studies (1999).

Table IV. Enhancing Orthene (Acephate) 755 Effectiveness against Cabbage Looper *Trichoplusiani*.

Treatment	*% Mortality of Cabbage Looper Larvae in Hours*			
	18	*44*	*68*	*92*
Orthene	0.2	50.0	82.6	91.7
Orthene + TPA	33.2*	69.4*	93.5*	98.8
Untreated (Check)	0.1	17.2	18.3	38.1

NOTE: * indicates significant differences at 0.05 level. Soil application of Orthene is 1 L/ha; label rate is 2 L/ha. TPA was applied at 5 L/ha.
SOURCE: Laboratory studies (1999).

The enhancement of insecticidal activity is likely due to the ability of TPA to enhance penetration (Table V) and residual of the active chemical (Table VI). Thermal polyaspartate can easily be mixed and applied with a wide range of insecticides, as given in Table VII.

Table V. Effect of TPA in Increasing Leaf Penetration of Neemix 4.5EC (Azadirachtin) against Greenhouse Whitefly *Trialeurodes vaporariorum*.

Treatment	*LC_{50} in PPM*	
	Lower Leaf	*Upper Leaf*
Neemix	56	35
Neemix + 1% TPA	30*	29

NOTE: * indicates significant differences at 0.05 level.
SOURCE: Greenhouse tests on cotton (1999).

Table VI. Effect of TPA in Enhancing the Residual Activity of Success (Spinosad) against Cabbage Looper *Trichoplusiani.*

Treatment	*LC_{50} in PPM*		
	24 HAT	*72 HAT*	*96 HAT*
Success	308	474	538
Success + 1% TPA	93*	226*	332*

NOTE: * indicates significant differences at 0.05 level.
SOURCE: Laboratory tests (sunlight chamber), 1999.

Table VII. Insecticides Compatible with TPA

Commercial Name	*Registered Trademark*	*Chemical Name*
Capture 2EC	FMC	Bifenthrin
Counter cr	BASF	Terbufos
Diazinon	Syngenta	Diazinon
Force 3G	Syngenta	Tefluthrin
Furadan	FMC	Carbofuran
Lorsban	Dow AgroSciences	Chlorpyrifos
Malathion	---	Malathion
Orthene	Valent	Acephate
Permit	Sanonda	Permethrin
Pounce	FMC	Permethrin
Sevin	Aventis	Carbaryl
SpinTor	Dow AgroSciences	Spinosad
Success	Dow AgroSciences	Spinosad
Zephyr	Syngenta	Abamectin

NOTE: List is based on physical compatibility tests and/or biological tests. Require continuous agitation during application.

Table VIII. Enhancing Herbicidal Effectiveness (Pre-Emergence Application) against Giant Foxtail and Velvetleaf.

Herbicide (0.25 Label Rate)	*% Biomass Reduction (g)*			
	Giant Foxtail		*Velvetleaf*	
	With TPA	*Without TPA*	*With TPA*	*Without TPA*
Harness	90.2*	45.0	90.2*	70.7
Dual	89.8*	61.1	---	---
Frontier	92.6*	78.3	---	---
Atrazine	97.7*	62.0	99.4*	29.7

NOTE: * indicates significant differences at 0.05 level. TPA applied at 5 L/ha. See Tables IX and X for scientific names of weeds.
SOURCE: Greenhouse studies *(17)*.

Table IX. Enhancing Roundup Original Effectiveness against Velvetleaf *Abutilon theophrasti.*

Roundup + TPA	*% Damage*	
	8 DAT	*16 DAT*
Label Rate	25.0 a	75.0 a
0.25 Label Rate	7.5 b	25.0 b
0.25 Label Rate + TPA	17.5 c	25.0 b
0.50 Label Rate	30.0 a	37.5 c
0.50 Label Rate + TPA	25.0 a	82.5 a

NOTE: Means followed by the same letters are not significantly different at the 0.05 level. Label rate is 0.8 kg/ha post-emergence application. TPA applied at 5 L/ha. DAT is days after treatment.
SOURCE: Greenhouse Studies, 1999.

Table X. Enhancing Roundup Original Effectiveness against Giant Foxtail *Setaria faberi.*

Roundup + TPA	*% Damage*	
	8 DAT	*16 DAT*
Label Rate	32.5 b	100.0 a
0.25 Label Rate	0.0 c	50.0 b
0.25 Label Rate + TPA	50.0 a	100.0 a
0.50 Label Rate	25.0 b	100.0 a
0.50 Label Rate + TPA	55.0 a	100.0 a

NOTE: Means followed by the same letters are not significantly different at 0.05 level. Label rate is 0.8 kg/ha post-emergence application: TPA applied at 5 L/ha. DAT is days after treatment.
SOURCE: Greenhouse Studies (1999).

Biological Data—Herbicides

The potential of TPA in enhancing the herbicides efficacy was tested in greenhouse and field environment. The results showed that the addition of TPA to various pre-emergence and post-emergence herbicides enhanced their effectiveness against a broad range of weeds. The use of TPA with lower-than-the-label-rate of certain herbicides gave equivalent or better control of the target weeds than their label rate. Also, in many cases, the addition of TPA to the label rate of certain herbicides enhanced their speed of action and/or effectiveness against certain weed species (Tables VIII through XI and Figures 1 through 3).

Thermal polyaspartate is also compatible with and easy to apply with a wide range of pre-emergence and post-emergence herbicides, as given in Table XII.

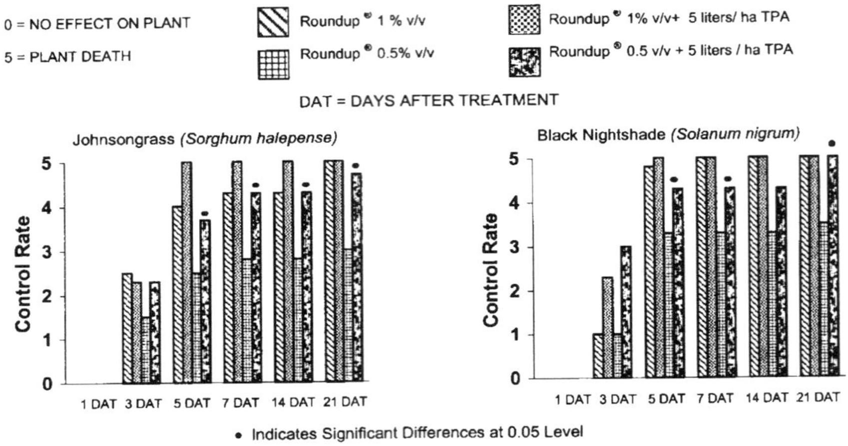

Figure 1. Enhancing herbicidal effectiveness of Roundup Original (post-emergence application). Field studies. California (1999).

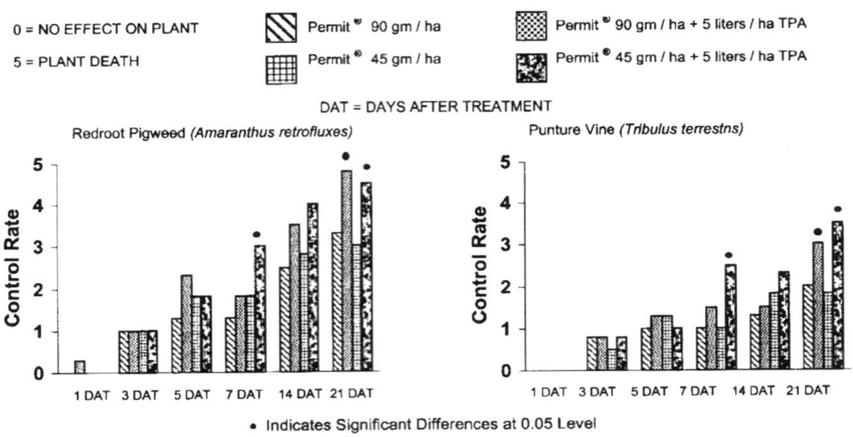

Figure 2. Enhancing herbicidal effectiveness of Permit (post-emergence application). Field Studies. California (1999).

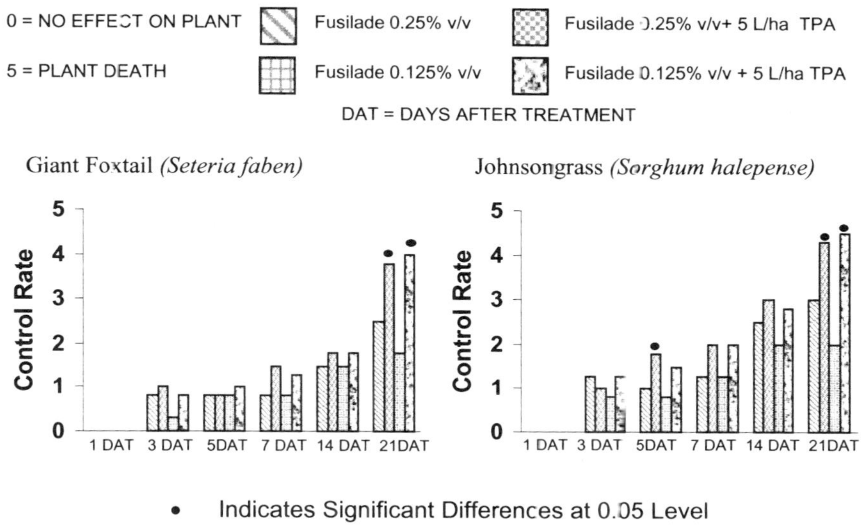

Figure 3: Enhancing herbicidal effectiveness of Fusilade (post-emergence application). Field studies. California (1999).

Table XI. Enhancing Herbicidal Effectiveness of Treflan HFP (Pre-Emergence Application).

Treflan + TPA (rates/ha)	*Mean % Control, 28 Days after Treatment*					
	Annual Morning glory	*Redroot Pigweed*	*Green Foxtail*	*Lamb-quarter*	*Barnyard grass*	*Total*
2.5 L	33.8	57.5	90.0	81.3	97.5	72.0
2.5 L + 2.5 L	70.0	67.0	93.3	82.0	94.0	81.3
2.5 L + 5.0 L	93.8	93.8	98.9	96.3	99.5	96.4
LSD (P=0.05)	18.9	20.8	16.4	16.5	11.3	12.8
S.D.	13.4	14.6	11.6	11.7	8.0	9.1

NOTE: Annual morningglory: *Ipomoea purpurea*, redroot pigweed: *Amaranthus refroflexus*, green foxtail: *Setaria viridis*, lambsquarter: *Chenopodium album*, barnyardgrass: *Echinochloa crusgalli*.

SOURCE: Field Studies. California (2000).

Table XII. Compatibility of TPA with Various Herbicides

Commercial Name	*Registered Trademark*	*Chemical Name*
	HIGH COMPATIBILITY	
Accord	Monsanto	Glyphosate
Classic	Dupont	Chlorimuron-ethyl
Hornet	Dow AgroSciences	Clopyralid + Flumetsulam
Lasso	Monsanto	Alachlor
Liberty	Aventis	Glufosinate
Pursuit	BASF	Imazethapyr
Python	Dow AgroSciences	Flumetsulam
Roundup Original	Monsanto	Glyphosate
Roundup Ultra	Monsanto	Glyphosate
	MEDIUM COMPATIBILITY	
Accent	Dupont	Nicosulfuron
Aim	FMC	Carfentrazone-ethyl
Axiom	Bayer	Metribuzin + Flufenacet
Balance	Aventis	Isoxaflutole
Basis Gold	Dupont	Atrazine + Rimsulfuron/ Nicosulfuron
Bicep II	Syngenta	Atrazine + Metolachlor
Buctril	Aventis	Bromoxynil Nitriles bromoxynil
Distint	BASF	Dicamba + Diflufenzopyr
Eradicane	Syngenta	S-ethyl dipropylthiocarbanate
Fusilade	Syngenta	Fluazifop-p-butyl
Fusion	Syngenta	Fenoxaprop-p-ethyl + Fluazifop-p-butyl
Gramoxone	Syngenta	Paraquat
Pinnacle	Dupont	Thifensulfuron-menthyl
Poast	BASF	Sethoxydim
Prowl	BASF	Pendimethalin
Raptor	BASF	Imazamox
Scepter	BASF	Imazaquin
Top Notch	Syngenta	Acetochlor
TouchDown	Syngenta	Sulfosate; glyphosate-trimesium
Treflan HFP	Dow AgroSciences	Trifluralin

NOTE: Based on physical compatibility tests. High compatibility requires no or minimum agitation during application; medium compatibility requires continuous agitation.

Conclusions

Laboratory, greenhouse, and field tests demonstrated that the addition of thermal polyaspartate to the label rate or lower-label rate of insecticides and herbicides enhanced their effectiveness against the target pests. The potential to optimize the uptake of pesticides promises to improve pest control, as well as to reduce the dissipation of chemicals into the environment. Additionally, TPA is known to enhance the ability of plants to uptake nutrients. These multiple benefits make TPA an ideal compound for use in agricultural crop management systems.

References

1. Low, K. C.; Wheeler, A. P.; Koskan, L. P. In *Hydrophylic Polymers: Performance with Environmental Acceptability*, Glass, J. E., Ed.; American Chemical Society: Washington, DC, 1996, 99–111.
2. Koskan, L. P.; Low, K. C. *U.S. Patent* 5,116,513, 1992.
3. Koskan, L. P.; Low, K. C.; Meah, A. R. Y.; Atencio, A. M. U.S. Patent 5,152,902, 1992.
4. Koskan, L. P.; Low, K. C. *U.S. Patent* 5, 284, 512, 1994.
5. McMahon, A. J.; Harrop, D. Green Corrosion Inhibitors: An Oil Company Perspective. *NACE CORROSION* 95 paper no. 32, 1995.
6. Jacquet, B.; Papantoniou, C.; Land, G.; Forestier, S. U.S. Patent 4,363,797, 1982.
7. Jacquet, B.; Papantoniou, C.; Land, G.; Forestier, S. *FR Demande* FR 2, 424, 292, 1979.
8. Kinnersley, A. P.; Koskan, L. P.; Strom, D. J.; Meah, A. R. Y. U.S. Patent 5,350,735, 1992.
9. Below, F. E.; Wang, X. T. *Agronomy Abstract*, Presented at ASA, CSSA, SSSA Annual Meetings, St. Louis, Missouri, 1995, p 107.
10. Murphy, L. S.; Sanders, J. L. Paper presented at ACS National Meeting, Division of Fertilizer and Soil Chemistry, American Chemical Society, Las Vegas, NV, September 7–11, 1997.
11. Gilbert, D. N.; Wood, C. A.; Kohlhepp, S. J., Kohnen, P. W.; Houghton, D. C.; Finkbeiner, H. C.; Lindsley, J.; Bennett, W. M. *J. Infect. Dis.* **1989,** *159,* 945–953.
12. Giammona, G.; Pitarresi, G.; Charlisi, B.; Cavallaro, G. J. *Bioact. Compat. Polym.* **1995,** *10,* 28–33.
13. Gleich, G. J. U.S. Patent 5,498,410, 1996.
14. Roweton, S.; Huang, S. J; Swift, G. *J. Environ. Polymer Degradation* **1997,** *5,* 175–182.

15. Tang, Y.; Wheeler, A. P. In *Biopolymers from Polysaccharides and Agroproteins*; Gross, R. A., Sholz, C., Eds.; ACS Symposium Series No. 786, American Chemical Society: Washington, DC, 2001, 157–171.
16. Sanders, J. L. U.S. Patent 5,709,890, 1998.
17. Sanders, J. L. U.S. Patent 5,635,447, 1997.

Chapter 9

Using Atom Transfer Radical Polymerization in Environmentally Benign Processes

Scott Gaynor, Jian Qiu, and Krzysztof Matyjaszewski

Department of Chemistry, Carnegie Mellon University, 4400 Fifth Avenue, Pittsburgh, PA 15213

Atom Transfer Radical Polymerization (ATRP) is a controlled/"living" radical polymerization system that allows for the synthesis of well-defined polymers with predetermined molecular weights, functionalities, and architectures. The ability to control these parameters enables the preparation of polymeric materials with novel, and heretofore unknown, utility. Materials prepared by ATRP perform more efficiently than comparable polymers, require less material to manufacture, and generate less catalytic waste than conventional processes. This paper describes the latest progress in the development of more efficient catalyst systems for ATRP, the extension to more environmentally friendly media, and some of the advanced materials prepared.

Introduction

Atom transfer radical polymerization (ATRP) is a metal-catalyzed process through which a reversible activation/deactivation of growing polymer chains suppresses termination and other chain breaking reactions (*1, 2*), Scheme 1.

ATRP allows for the preparation of smart materials that can be used more efficiently to perform the same application as other polymers, if they exist, but with less material. Additionally, the research that is being pursued is designed to use less transition metal catalyst, or to provide for economically feasible catalyst recovery, which prevents the generation of catalyst waste streams which would then require disposal.

The ATRP activation reaction involves the homolytic cleavage of a carbon-halogen bond (P_n–X) at the end of a polymeric chain (P_n) or a small molecule (R), that is, an initiator molecule, by oxidation of the transition metal catalyst (M_t^n/L). The moiety adjacent to the halogen stabilizes the radical, either through resonance or through induction. Some examples (and the polymers they mimic) include benzyl, 1-phenylethyl (styrenes), perfluoroalkyl (PTFE), 2-propionitrile (acrylonitrile), 2-propionate (acrylates), and 2-isobutyrate (methacrylates). Heteroatom-halogen bonds can also be activated and used for initiation, such as in sulfonyl chlorides (*3*).

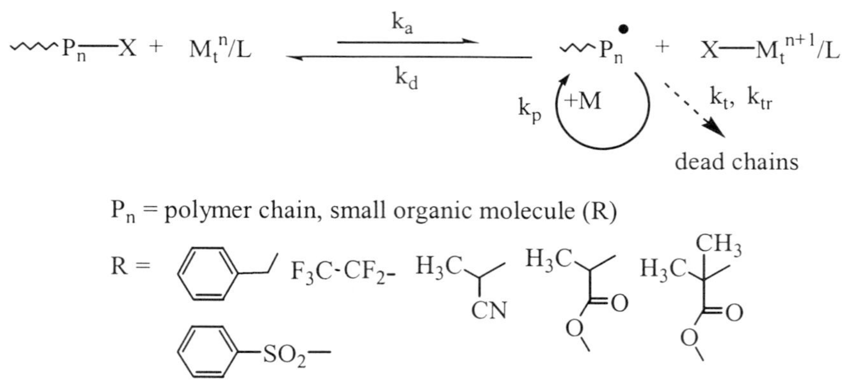

Scheme 1

The radical (R˙) formed as a result of the activation reaction follows three routes: initiates and continues propagation by addition of monomer (M); reacts with the transition metal halide (X-M_t^{n+1}/L) and reforms the dormant organic halide and the original catalyst (deactivation); or terminates via coupling or disproportionation. The fate of the radicals can be described by equation 1.

$$\frac{d[R\bullet]}{dt} = k_a[M_t^{\,n}/L][RX] - k_d[R\bullet][X - M_t^{\,n+1}/L] - k_t[R\bullet]^2 \qquad (1)$$

The stationary concentration of radicals is established by a balance between rates of activation and deactivation, rather than the balance between rates of termination and initiation, as in conventional radical polymerization. Through a process called the persistent radical effect (PRE) (*4*), an equilibrium between the radical and the dormant species is established in ATRP during the early stage of the polymerization, with the dormant species dominating (the stationary concentration of which is on the order of 10^{-2} to 10^{-1} M, versus 10^{-9} to 10^{-7} M for radicals). Because the reaction between the radical and the transition metal halide is designed to be so fast that only a few monomer units are added before the active radical is converted to the dormant alkyl halide, the polymer chains grow gradually and continually. The net result is the formation of well-defined polymer chains. These chains have the same end groups (from the R of the original initiator) and the halogen. The degree of polymerization (DP_n) of the polymer chain is described by the molar ratio of consumed monomer to added initiator ($DP_n = \Delta[M]/[R\text{-}X]_o$), and the distribution of molecular weights is relatively narrow (M_w/M_n <1.3). ATRP has been successfully applied to styrene (*5, 6*), acrylate (*7, 8*), methacrylate (*7, 9*), acrylonitrile (*10*), and acrylamide (*11, 12*) systems. Additionally, block copolymers have been prepared in various ways, including the addition of a second monomer after a first one is consumed (*6, 13*).

The challenge, however, is how to prepare polymers with fewer defects, in a shorter time, and with smaller amounts of catalyst. This requires a sound mechanistic understanding of the process so that faster and more versatile catalysts can be developed to facilitate the synthesis of advanced polymeric materials. The research at Carnegie Mellon University has been focused on these issues, and the rest of this publication will describe some of the progress in reducing the catalyst concentrations through catalyst development; the recovery/recycling of the catalyst from the final polymeric materials; some of the environmentally benign polymerization systems that ATRP has been conducted in; and materials that have been prepared using ATRP.

More Efficient ATRP Catalysts

One of the main drawbacks of ATRP is the relatively high amount of catalyst required in the reaction mixture, generally 0.1 to 1 percent. Therefore, significant efforts are made to remove the catalyst from the final product. To reduce the amount of metal complex used, it is desirable to improve the efficiency of the catalytic system. A better catalyst should form more radicals, meaning that they can be used at lower concentrations to provide the same rate of polymerization. In other words, a higher equilibrium constant, K_{eq}, between activation and deactivation of the growing radicals should be achieved. The increase of the equilibrium constant can be accomplished either by reducing the rate constant of deactivation (k_d) or by increasing the rate constant of activation (k_a). Because a fast rate of deactivation is desired for the preparation of well-defined polymer chains, the preferred pathway is to enhance the rate of activation. However, if the rate of activation is too fast (*14*), the equilibrium will be shifted strongly towards radicals, resulting in an uncontrolled polymerization with significant termination.

Changing the transition metal compound, ligand, additive, and/or the reaction medium has been an effective approach towards improving the catalytic system. When linear polyamines such as N,N,N',N",N"-pentamethyldiethylenetriamine and the corresponding heptamethyltriethylenetetramine (*15*) were used in place of bipyridines (*6*) and substituted bipyridines (*16*), the amount of the catalyst was reduced nearly 10-fold to achieve similar polymerization rates and molecular weight control. This resulted in a reduced cost of the ligand and reduced color of the reaction mixtures. The most efficient catalyst for ATRP discovered so far is a copper complex with tris[2-(dimethylamino)ethyl]amine (Me_6TREN) as the ligand. The polymerization of acrylate proceeds even at room temperature with less than 0.1 mol% catalyst (*17*). An example is shown in Scheme 2.

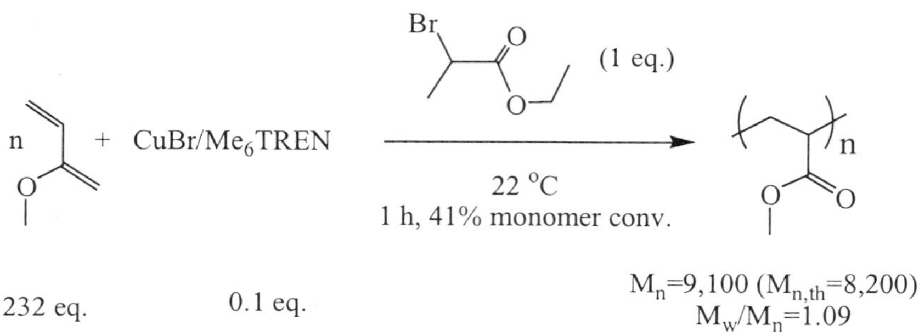

Scheme 2

The addition of a small amount of metal in the zero valency, such as Cu(0), also accelerates polymerization by reducing the corresponding Cu(II) species via an outer-sphere electron transfer (OSET) process. This enables a greater tolerance towards oxygen and radical inhibitors and also uses smaller amounts of ligands in the polymerization (*18*).

Recycling/Reclaiming of the ATRP Catalyst

Another approach to facilitate the removal and recycling of the catalyst is through immobilization. Attempts have been made to attach the catalyst to an insoluble support whereby the catalyst can react with the alkyl halide or radical. The general method is to prepare a support in which ligand groups are chemically bound to the support; the transition metal catalyst is then coordinated to the ligand and thus bound to the solid catalyst (*19-21*). While this method has been successful to some degree by providing polymers whose molecular weight increases with time, the polydispersities of the polymers were generally broad.

The reason for the broad molecular weight distributions has been attributed to the slow deactivation of the growing radical when in solution (*21*). Because the catalyst (deactivator) is not in solution but on the support, the growing polymer chain end must diffuse to the surface of the catalyst particle. This action is expected to be much slower than the diffusion of a small molecule in solution, which is how the unsupported catalyst would behave. The net result is that the rate of deactivation in the equilibrium between active and dormant species is drastically reduced (*vide supra*), thus providing a higher radical concentration and leading to increased termination and the addition of a larger number of monomer units per activation step.

To avoid this problem, Zhu and co-workers employed a soluble catalyst, Cu(I)Br/HMTETA (HMTETA = hexamethyltriethylenetetramine), but they adsorbed the catalyst onto a silica support (*22, 23*). Under appropriate conditions, they were able to prepare well-defined polymers, and the removal of the catalyst required only simple filtration of the silica. The catalyst could be recovered and reused numerous times with little loss in productivity and of polymerization control. The authors also published a report where they prepared packed columns of silica gel with the catalyst adsorbed onto it and used it successfully in repeated polymerization runs (*24*). These initial reactions were slow in comparison to the unsupported polymerizations (*15*), but the methodology holds promise and may provide an economic way of conducting ATRP on a commercial scale.

To retain the benefits of a polymerization using a soluble catalyst (faster rates of polymerization, well-defined polymers, etc.), methods were developed for efficient reclamation of the catalyst. Matyjaszewski and colleagues reported on the use of cationic ion exchange resins (*25*). The resins employed were crosslinked polystyrene beads containing $-SO_3H$ groups. The copper exchanged with the acidic proton and was chemisorbed onto the beads, leaving a clear solution. Haddleton et al. reported on the use of fluorous biphasic mixtures, where methyl methacrylate (MMA) was polymerized in perfluoromethyl cyclohexane using a fluorinated amine ligand for the copper catalyst (*26*). A controlled polymerization was obtained. Upon cooling of the reaction mixture, the organic and fluorous layers separated. Isolation of the organic layer and evaporation of the solvent yielded colorless polyMMA; the amount of copper found by ICP analysis in the polymer was 0.088% versus 1.5% if the catalyst had remained in the polymer.

Environmentally Friendly Reaction Media

Water

As a radical polymerization, ATRP can be conducted in the presence of water, thus reducing the VOC content of the polymerization system. Water has been used as a solvent for ATRP for the synthesis of water-soluble polymer. For example, 2-hydroxyethyl acrylate was polymerized in a 50 v/v% solution at 90 °C to afford a well-defined polymer (M_n = 14,700; M_w/M_n = 1.34) with 87% monomer conversion (*27*). Recently, Wang and Armes reported on the polymerization of oligo(ethylene glycol) methacrylate in water (*28*). Their results showed that well-defined polymers were obtained and that the reaction was faster in water than when conducted in the absence of water. Zhu and colleagues also reported the successful polymerization of 2-(dimethylamino)ethyl methacrylate in water (*29*).

For hydrophobic monomers, polymerizations in emulsions and suspensions were also successful (*30-32*). However, for the stabilization of insoluble polymer particles in an emulsion, it required that slightly modified conditions be used from those employed under homogeneous conditions. Original attempts at obtaining emulsions using conventional ionic surfactants, such as sodium dodecyl sulfate, resulted in polymerizations that were uncontrolled, that is, having ill-defined chain lengths and broad molecular weight distributions (*33, 34*). It was found that non-ionic surfactants bearing poly(ethylene oxide) moieties were capable of preparing well-defined polymers of styrene, acrylates and methacrylates, with narrow molecular weight distributions, M_w/M_n <1.3 (*33, 35*). Additionally, ATRP was later used to prepare well-defined polymers using conventional radical initiators (*36*) and also to synthesize block copolymers (*37*).

Carbon Dioxide

Another solvent system that has recently been developed as an environmentally-friendly alternative to organic solvents is carbon dioxide, either in its supercritical or liquid state (*38*). Although most polymers are insoluble in supercritical CO_2 ($scCO_2$), the precipitated polymer can be stabilized as discrete particles to prevent flocculation. This has been achieved by using block copolymer surfactants, where one block is soluble in the CO_2 phase and the other is compatible with the insoluble polymer (*39*). Venting of the CO_2 results in a free flowing powder.

The block copolymers that are employed generally contain a siloxane or perfluorocarbon polymer segment, along with a segment of an organic polymer, that is, styrene, methyl methacrylate. ATRP has been used as an effective method to prepare numerous block copolymers where the CO_2-soluble segment was a perfluoroacrylate and the insoluble segment a variety of acrylic and styrenic monomer types (*39, 40*).

ATRP was also conducted in $scCO_2$ but required the use of a special ligand for the copper catalyst. The catalyst was largely insoluble when conventional organic ligands such as 2,2'-bipyridine were used but was soluble when 4,4'-di(tridecafluoro-1,1,2,2,3,3-hexahydrononyl)-2,2'-bipyridine ($dR_{f6}bpy$) was added, turning the whole reaction mixture dark red. Homopolymers of 1,1-dihydroperfluorooctyl methacrylate (FOMA) and 1,1-dihydroperfluorooctyl acrylate (FOA) were prepared with predefined molecular weights, Scheme 3 (*40*).

The polyFOMA was then used as an initiator to prepare block copolymers with 2-(dimethylamino)ethyl methacrylate (DMAEMA) and MMA. The solubilities of the two copolymers were found to be different, with the FOMA-MMA being insoluble and the FOMA-DMAEMA giving cloudy solutions at 25 °C and 1800 psi. At higher temperature and pressure (55 °C, 5000 psi) the MMA copolymer formed a cloudy solution, while the DMAEMA copolymer was completely soluble.

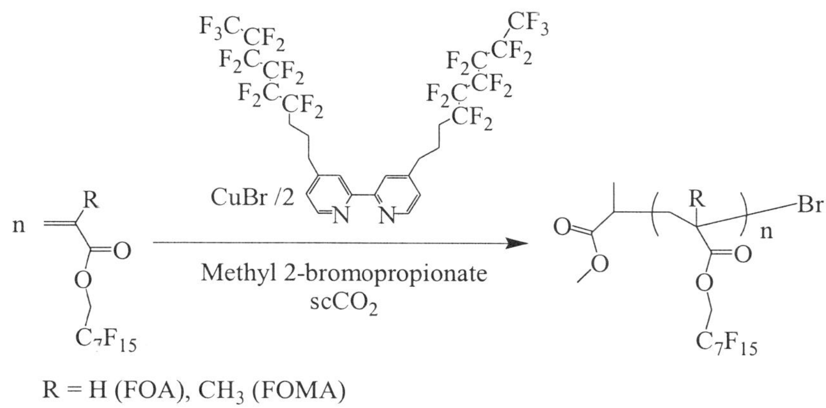

Scheme 3

Ionic Liquids

A new class of solvent media used for ATRP is ionic liquids. Haddleton and colleagues reported that 1-butyl-3-methylimidazolium hexafluorophosphate was successfully used as a polymerization medium for the synthesis of well-defined polyMMA (*41*).

Advanced Materials

Self-Plasticizing PVC

Poly(vinyl chloride) (PVC) is a tough and rigid material. While pure PVC has limited practical use due to its thermal instability and difficult processability, its great compatibility with a variety of additives and plasticizers significantly expands the potential applications of PVC. The plasticizers impart flexibility by serving effectively as internal lubricants that decrease the resistance between PVC chains, lowering both the melt viscosity and T_g, the glass-transition temperature. Low molecular weight plasticizers are commonly used in the production of plasticized PVC. However, they readily migrate to the surface of the product, which results in the deterioration of the physical properties of the material and in contamination of the environment. This drawback can be eliminated by grafting a rubbery polymer chain from PVC, hence producing a self-plasticizing material. ATRP has been used to graft poly(*n*-butyl acrylate) from PVC, as shown in Scheme 4 (*42*).

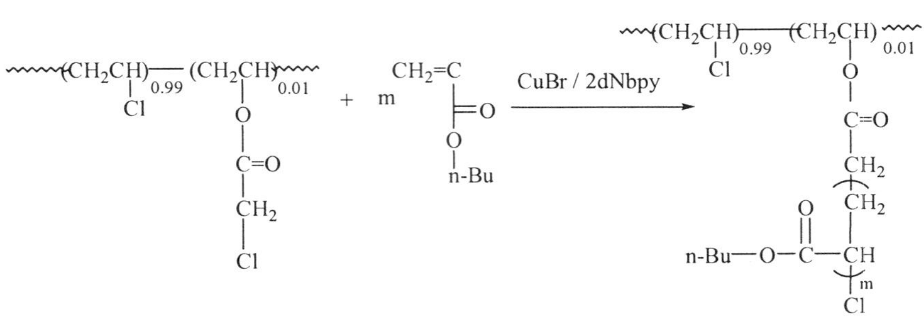

Scheme 4

The main chain incorporated 1 mol-% vinyl chloroacetate, which served as the ATRP initiator. The PVC backbone remained intact during ATRP, because the secondary chlorine on PVC could not be activated by the copper catalyst. The formation of the graft copolymer was confirmed by proton nuclear magnetic resonance (^{1}H NMR), Fourier transform infrared (FT-IR), and gel permeation chromatography (GPC). The films formed from the copolymers were transparent, indicating the absence of both macroscopic phase separation and large amounts of homopolymer formation. Thermal analysis (differential scanning calorimetry) revealed only one T_g for all copolymers, which decreased with increasing content of butyl acrylate incorporated (Table 1), as expected for the self-plasticizing material.

Table I. Results of Graft Copolymerization *n*-Butyl Acrylate from PVC

Time (hours)	M_n *(SEC)*	M_w / M_n *(SEC)*	M_n *(NMR)*	*Content of BA (mol-%)*	T_g*(°C)*
0	47400	2.66	-	0	83
2.0	61500	2.28	114,900	41	-4
4.3	79500	2.42	166,100	55	-11
9.5	81400	2.44	227,000	65	-19

NOTE: [Butyl Acrylate]$_o$ = 6.98 M, [CuBr]$_o$ = [dNbpy]$_o$ / 2 = 3.49 x 10^{-2} M, [PVC-1]$_o$ = 1.0 x 10^{-2} M, Temperature = 90 °C.

Acrylic Thermoplastic Elastomer

Thermoplastic elastomers (TPE) are segmented copolymers containing hard blocks (high T_g, such as polystyrene) and soft blocks (low T_g, such as butadiene), and therefore they display elastomer behavior around ambient temperature and thermoplastic behavior at typical processing temperatures. They are advantageous to chemically crosslinked rubbers in terms of processing and recycling. Compared with styrene-butadiene block copolymers, acrylic TPEs composed of methacrylates and acrylates are expected to have a greater resistance towards hydrocarbon solvents and to have improved optical properties.

The synthesis of block copolymers of methyl methacrylate (MMA) with either *n*-butyl acrylate (BA) or methyl acrylate (MA) was successfully achieved via copper mediated ATRP (*13*). The blocking sequence, however, has a great influence on the cross-propagation efficiency and hence requires particular attention. While PMMA generally provides good initiation efficiency for acrylate polymerization and forms clean block copolymers, polyacrylates are inefficient macroinitiators for the polymerization of MMA, due to the slower activation of acrylate-halogen bond relative to that of the methacrylate-halogen bond. This was improved by employing the halogen exchange technique, that is, using a bromine end group for the acrylate macroinitiator and CuCl as the catalyst. This allowed halogen exchange to occur; thus increasing the relative rate of initiation to propagation and efficiently yielding block copolymers with well-defined block lengths. Both di- and triblock copolymers have been obtained in various blocking sequences. The synthesis of acrylic block copolymers has recently been extended to the aqueous systems (*37*).

Non-Ionic Surfactants

Another advantage of ATRP is the facile synthesis of amphiphilic block copolymers, which can be used as nonionic surfactants and stabilizers for the aqueous dispersed polymerizations. With the molar mass, composition, and nature of the hydrophilic/hydrophobic blocks easily manipulated, these tailor-made surfactants are expected to attract substantial attention in both industry and academia. A variety of water-soluble monomers can be polymerized via ATRP (*27, 29, 43, 44*), forming amphiphilic block copolymers in combination with hydrophobic blocks (*45, 46*). These block copolymers are effective stabilizers in the free radical emulsion polymerization (*46*).

Solventless Coatings

Solventless coatings become increasingly important as industry moves toward systems with reduced VOC content. They are particularly attractive for applications where a short cure/drying time is desired, such as automotive coatings. In such areas the use of water-based systems becomes problematic because of the low volatility of water. While powder coatings have been widely used, their broad molecular weight distributions, originated from conventional radical polymerizations, result in non-uniform melting behavior of the powder. Additionally, the distribution in composition and functionality affects rheology and reactivity. The ability to flow under stress or elevated temperatures is important for processing and for obtaining a smooth coating surface. Uniform reactivity is desired in a coating material so that after curing, all polymer chains are a part of the uniform material. Chains that do not react may leach out over time, causing a change in the material properties, such as coating becoming brittle, hazy, and/or colored.

While all controlled polymerization methods allow for polymer chains to have nearly the same molecular weight and reactivity, ATRP is superior in controlling the functionalities, either at the end or in the center of the polymer chain. The former is accomplished by using functional initiators or by replacing the halogen end group. The latter can be approached by using multifunctional initiators. These strategies enable the preparation of various telechelics with identical and/or different functional groups, such as alcohol (*47*), epoxy (*48*), azide (*49–51*), amine (*48, 49*), or acid (*52*). These functional polymers, when prepared as α, ω-difunctional polymer chains, can react with a multi-functional curing agent to yield a cured material.

Other candidates for solventless coatings are highly functionalized hyperbranched polymers. Their synthesis via ATRP requires the use of a specially functionalized monomer, where the functional group is a latent initiator for ATRP, that is, an alkyl halide (*53–57*). These macromolecules have much smaller hydrodynamic volumes than their linear analogs with the same molecular weights. Therefore, they have lower melt viscosities and can be more easily processed. With appropriate functionality at the end of each branch, such as azide group, the polymer can be either irradiated or heated to form a crosslinked material (*58*). The functional groups can also participate in crosslinking reactions involving epoxides, isocyanates, or acid functionalized polymers.

Conclusions

ATRP is a very successful method for the preparation of a variety of polymeric materials. Many different research groups are currently developing new methods that will allow for ATRP to be conducted in an environmentally friendly and economical manner. Some of these methods involve polymerization in aqueous media or in CO_2, the use of more powerful or supported catalysts, and the facile reclamation of the catalyst.

Acknowledgments

The financial support from the Environmental Protection Agency's (EPA) Science To Achieve Results Program (STAR) is greatly appreciated. However, the research described in the article has not been subjected to any EPA review and therefore does not necessarily reflect the views of the Agency, and no official endorsement should be inferred.

References

1. Patten, T. E.; Matyjaszewski, K. *Adv. Mater.* **1998,** *10,* 901–915.

2. Patten, T.; Matyjaszewski, K. *Acc. Chem. Res.* **1999,** *32,* 895–903.
3. Percec, V.; Barboiu, B. *Macromolecules* **1995,** *28,* 7970–7972.
4. Fischer, H. *Macromolecules* **1997,** *30,* 5666–5672.
5. Patten, T. E.; Xia, J.; Abernathy, T.; Matyjaszewki, K. *Polym. Prepr. (Am. Chem. Soc., Div. Polym. Chem.)* **1996,** *37* (1), 575.
6. Wang, J.-S.; Matyjaszewski, K. *J. Am. Chem. Soc.* **1995,** *117,* 5614–5615.
7. Wang, J.-S.; Matyjaszewski, K. *Macromolecules* **1995,** *28,* 7901–7910.
8. Davis, K. A.; Paik, H.-J.; Matyjaszewski, K. *Macromolecules* **1999,** *32,* 1767–1776.
9. Grimaud, T.; Matyjaszewski, K. *Macromolecules* **1997,** *30,* 2216–2218.
10. Matyjaszewski, K.; Jo, S. M.; Paik, H.-J.; Gaynor, S. G. *Macromolecules* **1997,** *30,* 6398–6400.
11. Teodorescu, M.; Matyjaszewski, K. *Macromolecules* **1999,** *32,* 4826–4831.
12. Teodorescu, M.; Matyjaszewski, K. *Macromol. Rapid Commun.* **2000,** *21,* 190–194.
13. Shipp, D. A.; Wang, J.-L.; Matyjaszewski, K. *Macromolecules* **1998,** *31,* 8005–8008.
14. Queffelec, J.; Gaynor, S. G.; Matyjaszewski, K. *Macromolecules* **2000,** *33,* 8629–8639.
15. Xia, J.; Matyjaszewski, K. *Macromolecules* **1997,** *30,* 7697–7700.
16. Matyjaszewski, K.; Patten, T. E.; Xia, J. *J. Am. Chem. Soc.* **1997,** *119,* 674–680.
17. Xia, J.; Gaynor, S. G.; Matyjaszewski, K. *Macromolecules* **1998,** *31,* 5958–5959.
18. Matyjaszewski, K.; Coca, S.; Gaynor, S. G.; Wei, M.; Woodworth, B. E. *Macromolecules* **1997,** *30,* 7348–7350.
19. Haddleton, D. M.; Duncalf, D. J.; Kukulj, D.; Radigue, A. P. *Macromolecules* **1999,** *32,* 4769–4775.
20. Haddleton, D. M.; Kukulj, D.; Radigue, A. P. *Chem. Commun.* **1999,** 99–100.
21. Kickelbick, G.; Paik, H.-J.; Matyjaszewski, K. *Macromolecules* **1999,** *32,* 2941–2947.
22. Shen, Y.; Zhu, S.; Zeng, F.; Pelton, R. *Macromol. Chem. Phys.* **2000,** *201,* 1387–1394.
23. Shen, Y.; Zhu, S.; Zeng, F.; Pelton, R. H. *Macromolecules* **2000,** *33,* 5427–5431.
24. Shen, Y.; Zhu, S.; Pelton, R. *Macromol. Rapid Commun.* **2000,** *21,* 956.
25. Matyjaszewski, K.; Pintauer, T.; Gaynor, S. *Macromolecules* **2000,** *33,* 1476–1478.
26. Haddleton, D. M.; Jackson, S. G.; Bon, S. A. F. *J. Am. Chem. Soc.* **2000,** *122,* 1542–1543.

27. Coca, S.; Jasieczek, C.; Beers, K. L.; Matyjaszewski, K. *J. Polym. Sci., A: Polym. Chem.* **1998,** *36*, 1417–1424.
28. Wang, X. S.; Armes, S. P. *Macromolecules* **2000,** *33*, 6640–6647.
29. Zeng, F.; Shen, Y.; Zhu, S.; Pelton, R. *J. Polym. Sci., A: Polym. Chem.* **2000,** *38*, 3821–3827.
30. Qiu, J.; Pintauer, T.; Gaynor, S. G.; Matyjaszewski, K.; Charleux, B.; Vairon, J.-P. *Macromolecules* **2000,** *33*, 7310–7320.
31. Matyjaszewski, K.; Qiu, J.; Tsarevsky, N.; Charleux, B. *J. Polym. Sci., A: Polym. Chem.* **2000,** *38*, 4724–4734.
32. Wan, X.; Ying, S. *J. Appl. Polym. Sci.* **2000,** *75*, 802–807.
33. Gaynor, S. G.; Qiu, J.; Matyjaszewski, K. *Macromolecules* **1998,** *31*, 5951–5954.
34. Makino, T.; Tokunaga, E.; Hogen-Esch, T. E. *Polym. Prepr. (Am. Chem. Soc., Div. Polym. Chem.)* **1998,** *39* (1), 288–289.
35. Jousset, S.; Qiu, J.; Matyjaszewski, K. *Macromolecules* [Online] **2001,** *34,* ASAP Contents, Web Release Date: Aug 15, 2001.
36. Qiu, J.; Gaynor, S. G.; Matyjaszewski, K. *Macromolecules* **1999,** *32*, 2872–2875.
37. Matyjaszewski, K.; Shipp, D. A.; Qiu, J.; Gaynor, S. G. *Macromolecules* **2000,** *33*, 2296–2298.
38. DeSimone, J. M.; Guan, Z.; Elsbernd, C. S. *Science* **1992,** *257*, 945.
39. DeSimone, J. M.; Betts, D.; Johnson, T.; McClain, J. M.; Wells, S. L.; Dobrynin, A.; Rubinstein, M.; Londono, D.; Wignall, G.; Triolo, R. *Polym. Prepr. (Am. Chem. Soc., Div. Polym. Chem.)* **1999,** *40* (1), 435–436.
40. Xia, J.; Johnson, T.; Gaynor, S. G.; Matyjaszewski, K.; DeSimone, J. M. *Macromolecules* **1999,** *32,* 4802–4805.
41. Carmichael, A. J.; Haddleton, D. M.; Bon, S. A. F.; Seddon, K. R. *J. Chem. Soc., Chem. Commun.* **2000,** 1237–1238.
42. Paik, H.-J.; Gaynor, S. G.; Matyjaszewski, K. *Macromol. Rapid Commun.* **1998,** *19*, 47–52.
43. Xia, J.; Zhang, X.; Matyjaszewski, K. *Macromolecules* **1999,** *32,* 3531–3533.
44. Zhang, X.; Xia, J.; Matyjaszewski, K. *Macromolecules* **1998,** *31,* 5167–5169.
45. Davis, K. A.; Charleux, B.; Matyjaszewski, K. *J. Polym. Sci. A: Polym. Chem.* **2000,** *38,* 2274–2283.
46. Burguiere, C.; Pascual, S.; Coutin, B.; Polton, A.; Tardi, M.; Charleux, B.; Matyjaszewski, K.; Vairon, J.-P. *Macromol. Symp.* **2000,** *150,* 39–44.
47. Coessens, V.; Matyjaszewski, K. *Macromol. Rapid Commun.* **1999,** *20*, 127–134.
48. Matyjaszewski, K.; Coessens, V.; Nakagawa, Y.; Xia, J.; Qiu, J.; Gaynor, S. G.; Coca, S.; Jasieczek, C. *ACS Symp. Ser* **1998,** *704*, 16–27.

49. Nakagawa, Y.; Gaynor, S. G.; Matyjaszewski, K. *Polym. Prepr. (Am. Chem. Soc., Div. Polym. Chem.)* **1996,** *37* (1), 577–578.
50. Coessens, V.; Nakagawa, Y.; Matyjaszewski, K. *Polym. Bull.* **1998,** *40,* 135–142.
51. Coessens, V.; Matyjaszewski, K. *J. Macromol. Sci.* **1999,** *A36* (5 & 6), 667–679.
52. Zhang, X.; Matyjaszewski, K. *Macromolecules* **1999,** *32*, 7349–7353.
53. Gaynor, S. G.; Edelman, S. Z.; Kulfan, A.; Matyjaszewski, K. *Polym. Prepr. (Am. Chem. Soc., Div. Polym. Chem.)* **1996,** *37* (2), 413–414.
54. Hawker, C. J.; Frechet, J. M. J.; Grubbs, R. B.; Dao, J. *J. Am. Chem. Soc.* **1995,** *117,* 10763–10764.
55. Gaynor, S. G.; Matyjaszewski, K.; Muller, A. H. E. *Macromolecules* **1997,** *30,* 7034–7041.
56. Gaynor, S. G.; Kulfan, A.; Podwika, M.; Matyjaszewski, K. *Macromolecules* **1997,** *30,* 5192–5194.
57. Matyjaszewski, K.; Pyun, J.; Gaynor, S. G. *Macromol. Rapid Commun.* **1998,** *19,* 665–670.
58. Gaynor, S. G.; Matyjaszewski, K. *ACS Symp. Ser.* **1998,** *685,* 396.

Chapter 10

Generating Benign Alternative Syntheses: The SynGen Program

James B. Hendrickson

Department of Chemistry, Brandeis University, Waltham, MA 02454

A rigorous system of synthesis design is presented to find all the best syntheses for any target, based on skeletal dissection and a radical simplification in characterizing compounds and reactions. The SynGen program to execute this design model is described with examples.

Introduction

A broad view of the chemical industry suggests two general approaches to improving the environmental impact of any industrial synthesis. The first and more apparent is to improve the existing synthesis by improving yields and substituting less hazardous reagents. The second approach is to examine completely different, alternative synthetic plans, seeking shorter pathways and more benign input chemicals. This second approach, addressed here, is less common for two reasons. Having devoted much time to its development, chemists become quite committed to the current synthesis route and are loath to believe it possible that any quite different alternative could be better. The second reason that such alternative routes are rarely sought is that there has traditionally not been any systematic procedure to discover all of them.

System for Synthesis Design

For most target molecules there are indeed many thousands of potential synthesis pathways, more than are generally appreciated. For a number of years our group has focused on the problem of synthesis design, specifically of formulating a rigorous protocol to locate the best synthetic routes to any desired target from so many possible ones (*1*). Such a system must be based on clear criteria and proceed independently, without any user bias. Our design system has led to the program SynGen for generating the best syntheses of any target molecule. The logical basis of the system is outlined here. The labeled sections below develop the main features of the design logic, and these are then encapsulated in five main procedural steps in italics.

Criteria

As an initial guide through the morass of possible routes, we must first establish stringent criteria for what constitutes the best syntheses. We settled on a key demand that the shortest syntheses are the best. This takes into account the fact that most reactions in any route are imperfect and lead to byproducts, often environmental contaminants, which must be disposed of. Further, any reaction consumes time, reagents, and solvents, and so the fewest reaction steps must result in the least contamination and lowest cost.

To achieve the fewest steps in a synthesis route also requires starting materials that are structurally close to the target, requiring the least change on incorporating them into the target molecule. This in turn means that we need a large and suitably organized catalog of starting materials. Additionally, environmental concerns demand that the best routes will use starting materials that are not toxic or carcinogenic, nor show other environmental hazards.

Minimizing waste also demands knowledge of the yield of each reaction. Accurate yield prediction of new, untried reactions generally is not possible, but we can at least compare proposed reactions against a large database of closely matching reactions from the literature to find an average yield. With these average yields and the starting material prices, we can calculate an overall cost for any route and so order the generated routes by their cost.

1. *Define the best routes as those with the fewest steps, the lowest cost, and minimal environmental hazard.*

Skeleton

Our analysis of synthesis generation starts by simplifying the problem, beginning with just the skeletons of the molecules. A fundamental observation about most syntheses is that a large target is made from small starting materials. As a consequence, it is obligatory that some skeletal bonds of the target must be constructed in any synthesis. This idea allows us to envision any target skeleton as an assembly of linked starting material skeletons. In fact, it is easy to generate all possible skeletal assemblies of the target from different sets of starting material skeletons. To generate routes retrosynthetically, we must first cut the target skeletal bonds that link these starting material units. Thus in the forward direction the central reactions of the synthesis routes will be the construction reactions that create those linking bonds, mainly C–C bonds.

We start by cutting the target skeleton in two in the search for smaller fragments, and we do this all possible ways. This is done either by cutting one acyclic bond or two exterior bonds in any one ring in the target skeleton. In the forward direction this ring-cutting corresponds to an annelation reaction, which joins two molecules first at one skeletal bond and then cyclizes, forming a second bond in the ring. This is generally faster and easier so that often both bonds may be made in one step, which is advantageous.

With the starting materials catalog organized by their skeletons, we can discover if either of the cut pieces has the skeleton of an available starting material. If not, each piece may be divided again in all possible ways, and the smaller skeletal fragments so derived can be compared again with the catalog.

There will usually be many ways to cut any skeleton into two pieces. These are then ordered so as to present the most equal-sized skeletal fragments first, as these tend to be more likely to produce the shortest and most convergent overall routes in the forward direction.

2. *Cut the skeleton into two pieces, cutting one acyclic bond or two ring bonds. Repeat the process, seeking catalog skeletons.*

Functionality

Every skeletal bond cut must be made in the forward direction by a construction reaction. To satisfy the criterion of fewest steps, we demand that *every* reaction in a synthesis route must be a construction of a cut skeletal bond. Hence no other reactions, for example, to repair functional groups, are allowed between the constructions. Such routes, of consecutive constructions only, must then be the shortest possible routes and are described as "ideal syntheses". An ideal synthesis requires that the last construction in any route must make not only the last skeletal bond but also the correct target functionality. This functionality fit will be required of each preceding construction as well. Thus in an ideal synthesis the chosen starting materials are so functionalized as to pass through a sequence of construction reactions to arrive directly at the target, not only with its skeleton but also with its full functionality in place.

Knowing the functional groups of the target around the last skeletal bond to make, we can retro-generate those construction reactions that can form both that skeletal bond and the correct functional groups around it. When two cuts in a ring are required, this is done for both bonds in either order. This process then generates two smaller precursors, now with defined functional groups to initiate the required constructions. These precursors are compared with the starting material catalog to find out whether they are available starting materials. If not, they become in turn targets for repeating the process.

> 3. *Retro-generate the construction reactions for the cut bonds that create the target. Compare the smaller precursors so generated with the starting material catalog. Repeat the whole process, generating yet smaller precursors, until catalog compounds are found for each.*

The first level of cuts divides the original target into two smaller precursors. At the second level one or both of these is cut in two again, so that up to four final starting compounds will have been generated in two levels. Only if all of these can be found as starting materials in the catalog will we record it as an acceptable synthesis route. In our experience two levels of cuts generally suffice to generate an adequate number of synthesis routes from real starting materials, and these must be the shortest possible "ideal" syntheses.

The routes may then be ordered by number of steps and/or overall cost. If not enough routes have been generated, one or more of these second-level precursors may be cut in two again for a third level. Syntheses taken to three levels will not only be longer and costlier, but also in general will require much more time and will create far too many routes.

> 4. *Limit the number of cut levels to two for the best routes, only going to a third level if too few routes have been found.*

Reactions

To encompass the huge amount of detail implicit in these reactions and to coalesce trivial distinctions, the description of structures and reactions must be stringently abstracted and generalized. To do this we define only four general kinds of bonds to carbon, and so each carbon is described digitally by the number of each kind of bond. For the computer, any compound may therefore be generalized as a binary number, that is, an ordered list of its skeletal carbons, each digitally defined by these bond types.

Of the four defined kinds of bond, the first is a σ-bond to another carbon. This defines the skeleton, showing the number of σ-bonded carbons attached to any carbon. The π-bonds to adjacent carbons are defined separately, because they are functional, as are the other two kinds of bonds: to electronegative atoms (N, O, X, S, P) and to electropositive atoms (H or metals). In this way the system generalizes families of functional groups that have roughly the same kind of reactivity. This abstraction to these four kinds of bonds also affords calculation of oxidation state and its change in any reaction.

In these generalized terms the construction reactions, which make the skeletal σ-bonds, involve creating a new σ-bond between two carbons at the expense of another kind of bond at each one. All combinations of such bond type exchanges can be defined in advance to create a complete roster of all possible kinds of net structural change in construction reactions. It is this roster that is applied to the carbons at the reaction center to generate retrosynthetically all the precursor functionalities for any construction of a defined skeletal bond and its associated functional groups.

When precursor compounds are thus defined for any construction, their comparison with the catalog compounds, similarly generalized, is a simple and rapid comparison of their defining binary numbers by the computer. Overall, this process affords a series of routes from a full set of available starting materials, with the reactions leading from them to the target, usually in just two levels of constructions and requiring at most only six reaction steps.

5. *To encompass all possibilities rapidly, generalize the structures and reactions first, and only refine them to full chemical detail at the end after the few best routes have been selected.*

Reaction Validation

Each reaction generated may be compared with the literature. Its reaction center is matched with a table of similarly generalized entries from a large reaction database. The matching uses the WebReactions fast retrieval system (*2*) from a database winnowed to construction reactions only (about 100,000 entries from 1975 to 1995). For any reaction, the table shows the frequency and average yield of matching constructions in the database. The average yield is then used to calculate the overall yield and cost of any route. This look-up of matches from the literature serves as a measure of confidence for any generated reaction. Detailed matching database entries may be sought separately from http://www.WebReactions.net (*3*).

Refunctionalization

Although generally the procedure requires that all reactions be construction reactions, there is one place where altering the functional groups on the same skeleton can be useful without compromising the stringent demands of the "ideal synthesis". The catalog can never be as large as is needed to generate all routes, and so it is in effect expanded by allowing a simple one-step functionality change on the same skeleton. Thus when the program generation of precursors finds no match in the catalog, it then seeks functional variants of the catalog entries with the same skeleton but only one step away, and these are then incorporated into the final routes, which are acceptable but one step longer than "ideal".

The SynGen Program

Synthesis Generation Procedure

The central module of the program proceeds from the target, or any product, and cuts its skeleton in two in all possible ways. Then for each such cut it derives retrosynthetically all possible construction reactions that form the product functionality around the cut bond. This generates the functionalized precursors, and they are compared with the catalog compounds. The catalog contains about 25,000 commercially available compounds. Any precursor not in the catalog is returned to the central module as a product to be cut in two again and its precursors compared with the catalog. When two ring bonds in a skeleton must be cut, they are sequentially examined in both orders to yield pairs of precursors.

SynGen passes through all first level generations from the target, generating all pairs of intermediates. It then proceeds through the second level generations of all precursors of each of those intermediates. Normally the generation of chemistry stops here, with a set of whole routes complete in two levels and all precursors found in the catalog of starting materials. If a second level precursor is not found, its preparation in one step by refunctionalizing a catalog compound is sought. At this stage the overall cost per mole of the target is calculated from the starting material prices and the average predicted yields of each reaction, and the set of routes is ready for display, ordered by cost and/or by number of steps.

If enough whole routes are not created in two levels, the user may elect to cut again, for a third level, no more than two second-level precursors not found in the catalog. This takes much longer and often produces too many routes.

In practice, the user enters the structure of a target molecule, saved in Molfile format from an external drawing program. The user may elect some prior limitations on the allowed skeletal cutting. Then SynGen takes over and quickly generates whole, "ideal synthesis" routes that begin from available starting materials, usually in less than five minutes. The program may be downloaded from the web at http://syngen2.chem.brandeis.edu/syngen.html.

The user may then scan this output as outlined in the next section. It is common to find that a very large number of routes may have been generated, even with the minimalist demands of these criteria. This is a consequence of the mechanical nature of computer generation, but navigation among the routes is easy and will soon elucidate the best ones, as described below.

Output Displays

When the generation is complete, the program offers a window of viewing choices, the first of which is the "Target" display, summarizing the generation of the number of successful routes obtained, the time required, and the conditions set by the user at the beginning. All of this is apparent in Figure 1, a summary of the 775 routes found for lysergic acid, the parent molecule of the ergot alkaloid medicinal drugs (*4*).

There are three other kinds of displays for flexible examination of the results. The first of these, labeled as "First-level Bondsets", shows the successful skeletal cuts of the target, as in Figure 2 for lysergic acid. The user may select any one or more of these bondsets for examination; in the figure one bondset of 256 routes has been temporarily deleted. These bondsets are identified by letters at the top, two letters for two ring cuts indicating which of the bonds is made first. In the figure both orders *D* and *E* were deleted; in the first bondset, for the two cuts labeled *A* only one order afforded any successful routes.

The central viewing choice is the "Full Route" display, which shows all the routes one at a time via the navigation arrows (forward/backward) at left on the toolbar. The routes are ordered either by overall cost or by number of steps, and

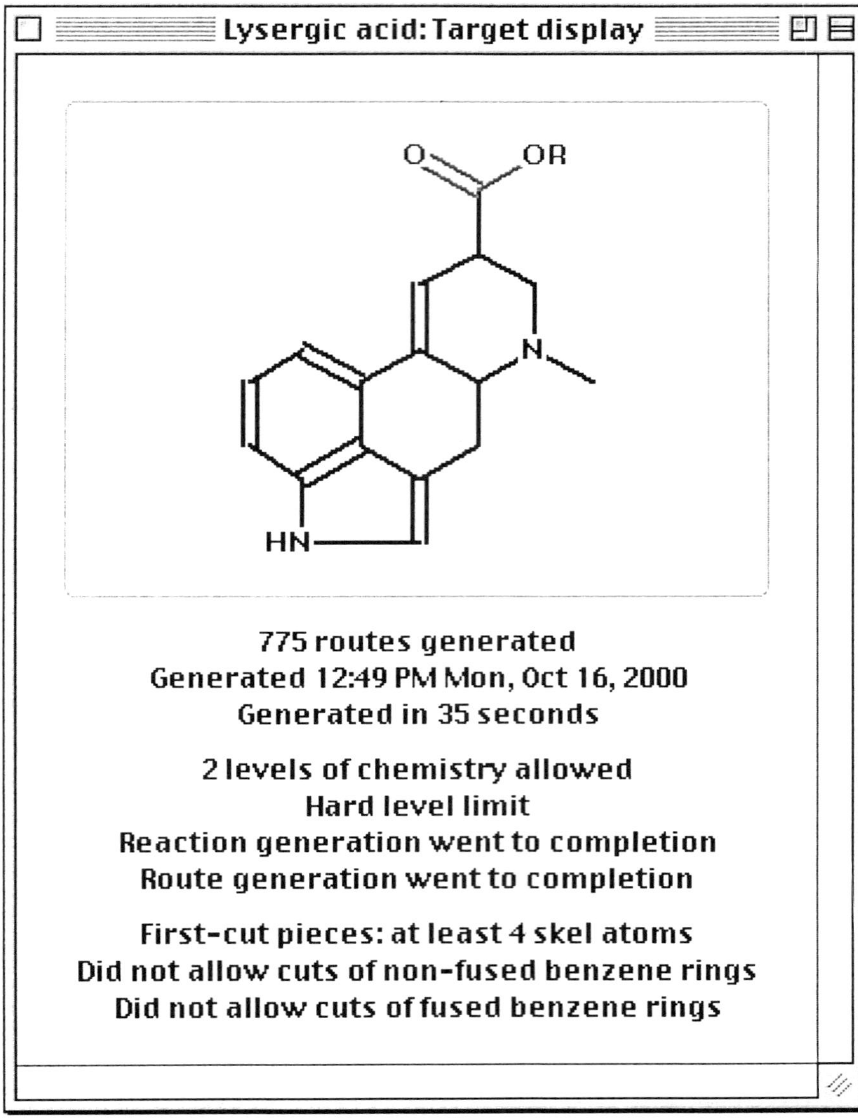

Figure 1. Display screen for target and generation conditions.

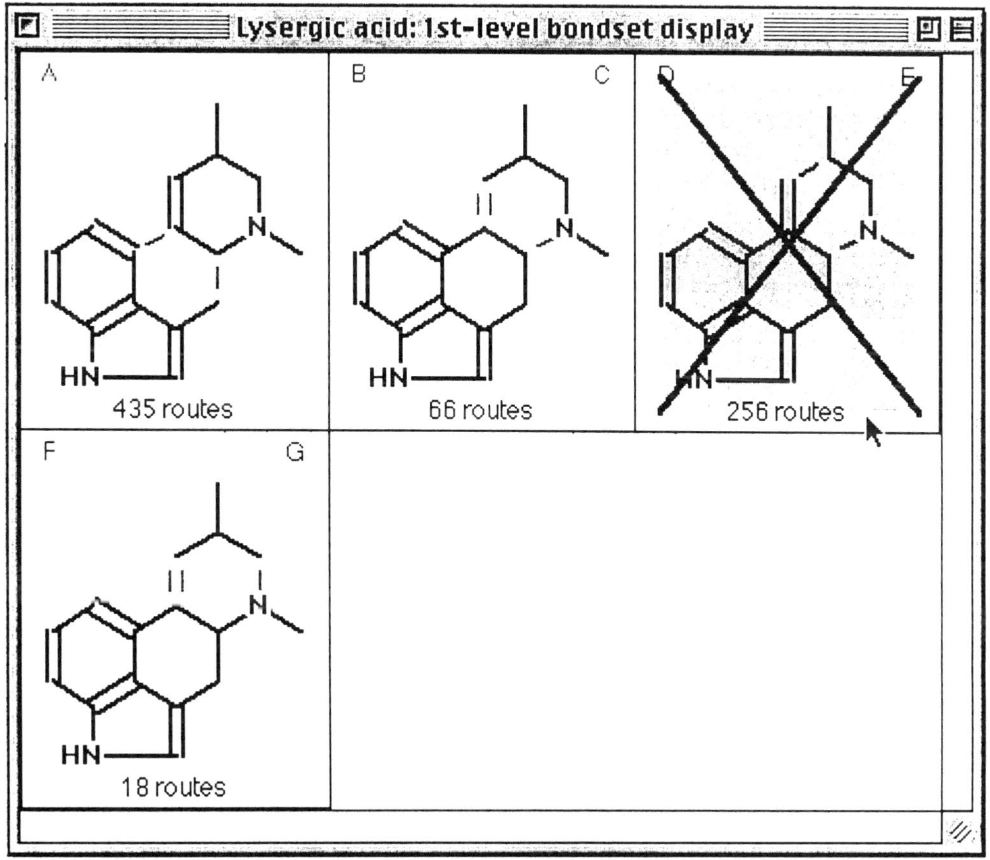

Figure 2. Bondset display with one deleted.

the routes are identified on the toolbar by bondset (letter) and number. Figure 3 shows the cheapest route (C1) of the 519 remaining after the bondset deletion of Figure 2.

The nature of each reaction and its database background may be shown in a window by clicking on any reaction arrow. Clicking on any starting material structure shows its price per mole, its supplier, and most importantly its environmental hazard on a uniform logarithmic scale.

The third viewing choice is the "Build Route" display. Because of the expanding nature of the generation of routes back from the target, the number of last reactions to the target, joining the first-level intermediates, is relatively small. In this display the user may see all of these last reactions and choose to explore any one of them backwards, clicking on each intermediate in turn to see how many reactions form it, and so for each choice working back to find the starting materials. This procedure is a flexible method of probing for the most viable syntheses and is examined further in the next section.

The "Edit" menu (not shown in the figures) also provides for flexible deletion and retention options as well as providing a preferences file for deleting routes with functional group interferences or other questionable reactions perceived by the program.

Navigation: Search for Optimal Routes

The user will probably get many synthetic routes because of the mechanical nature of their generation, affording many reactions that are just minor variations. The numbers of routes are also compounded because, if two intermediates are to be joined, the variations in ways to make one must all be applied to each variation in the ways to make the other.

However, large numbers of routes are not a problem with some simple navigation direction to pick out the main themes and to quickly afford a structured view of the output over the many small and repeated variations in the chemistry that were generated by the program.

The "First-level Bondset" display shows the several dissections of the target skeleton into two intermediate skeletons. As such, it allows an initial focus, as in Figure 2, on one or more target assemblies, and each of these carries implications about the kinds of starting materials. As an example, if one were interested in employing tryptophan as a starting material for lysergic acid, then the whole skeleton of tryptophan must be intact in one of the two intermediates. This is only true of the fourth bondset (F/G) in Figure 2, and so selecting only this one affords only 18 routes. The cheapest and shortest of these is shown in Figure 4 as route F2. The tryptophan starting material is shaded here. The

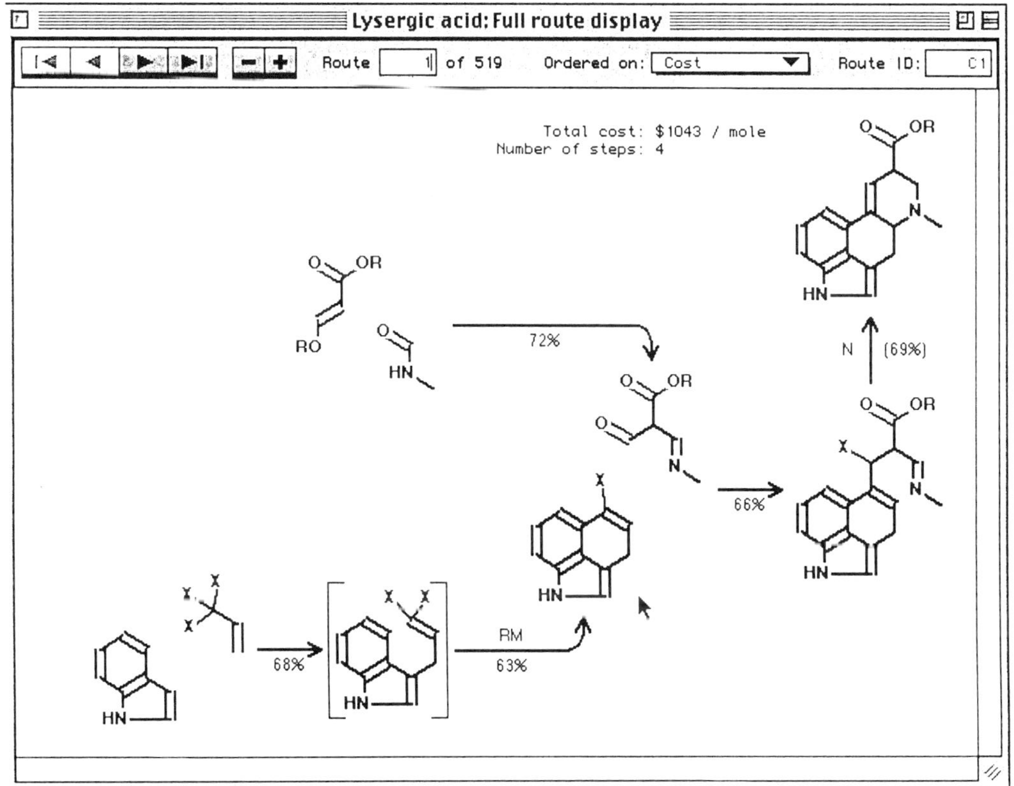

Figure 3. Full route display for least costly route.

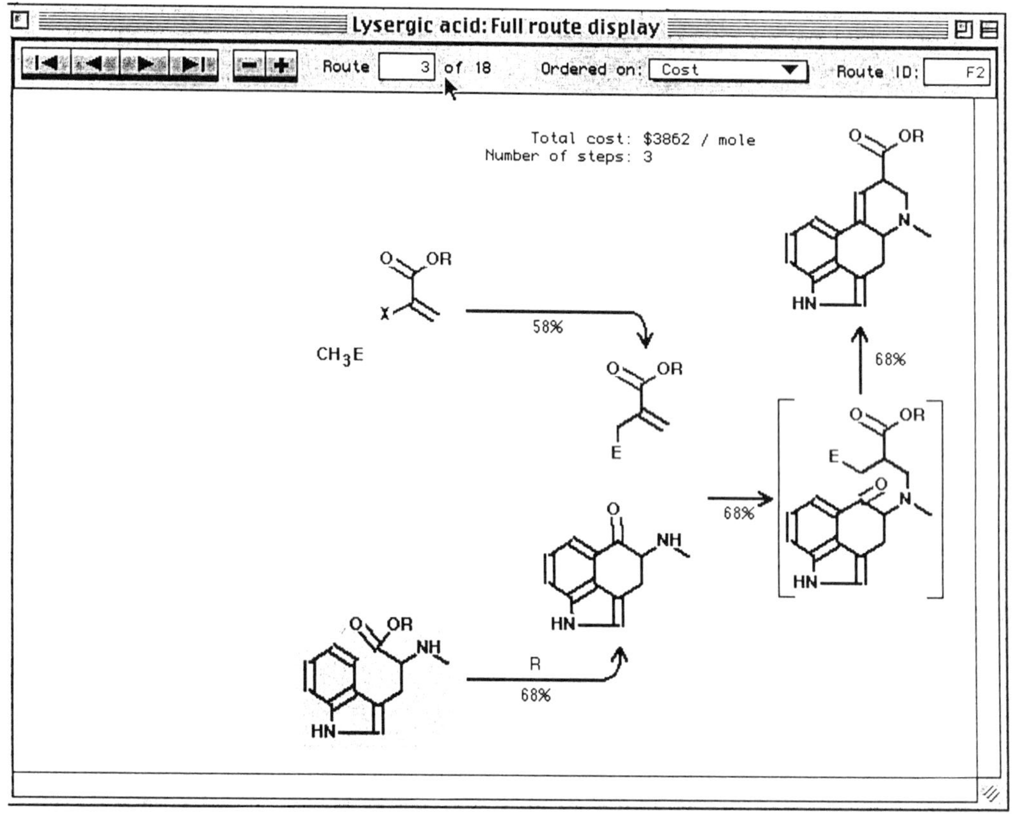

Figure 4. Full route starting with tryptophan.

generalized substituent -E is used for an electron-withdrawing group in the other compounds. Route F2 is the basis of an actual synthesis of lysergic acid (*4*).

The "Build Route" display is the most versatile, allowing successive choices at each step back from the target. The display begins with all of the last reactions to the target, either from the whole set of routes or just within the subset from any selected bondset(s). The toolbar shows how many last reactions there are, and these may be examined one at a time with the navigation arrows on the toolbar. Figure 5 shows the 13th of 24 last reactions from the full set of routes to lysergic acid.

Having selected a favored last reaction, the user clicks on its precursor to see all ways to form that precursor. Figure 6 shows the precursor of Figure 5 selected (shaded) and displays the third of three reactions to form it. This is continued by selecting separately the two intermediates shown in Figure 6. The upper one is a generalized starting material actually available as its nitrile, shown as the actual catalog starting material in the window. The lower intermediate in Figure 7 affords two syntheses, one of which is shown in the figure, which is now a complete route labeled as E8 in the upper right corner.

This procedure allows the user to make his own choice of the best reactions to form each intermediate, and so the user easily builds up favored whole routes from the large set originally generated. These favored whole routes may then be retained for later display or printing.

The "Full Route" display may also be used, with or without a prior selection from the "Bondset" display, by starting with the best (in terms of cost or number of steps) full route and then clicking through the following ones. Each time any intermediate or reaction in a route appears to be not viable, it may simply be deleted by clicking on the item. Once deleted, any selected compound or reaction will not be displayed again in any other variant route. When any item is deleted, the total number of routes shown on the toolbar is then diminished, often substantially, indicating how many other routes had also included that deleted item.

This reduction in the number of routes can be substantial, and a few such deletions can quickly establish a smaller and more viable set of remaining routes to consider. Interesting routes may at any time be marked for retention (in the "Edit" menu, not shown) for later viewing or printing. As an example, take the best of the 519 routes shown in Figure 3 and delete the tricyclic intermediate marked with the cursor arrow. When this is done only 511 routes remain. An interesting new route now appears here (Figure 8) as the best of those remaining, somewhat more costly (arrow) but with only three steps, that is, route A1 with a quite different skeletal assembly from that in Figure 3.

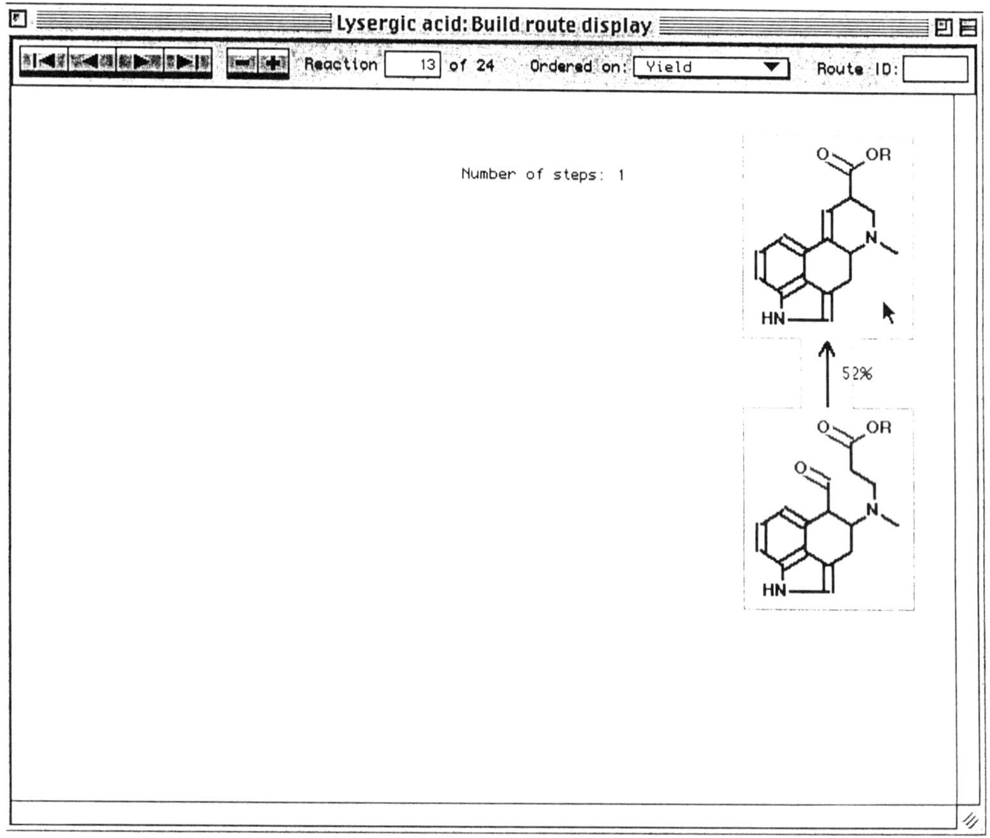

Figure 5. Build route display for last reaction to target.

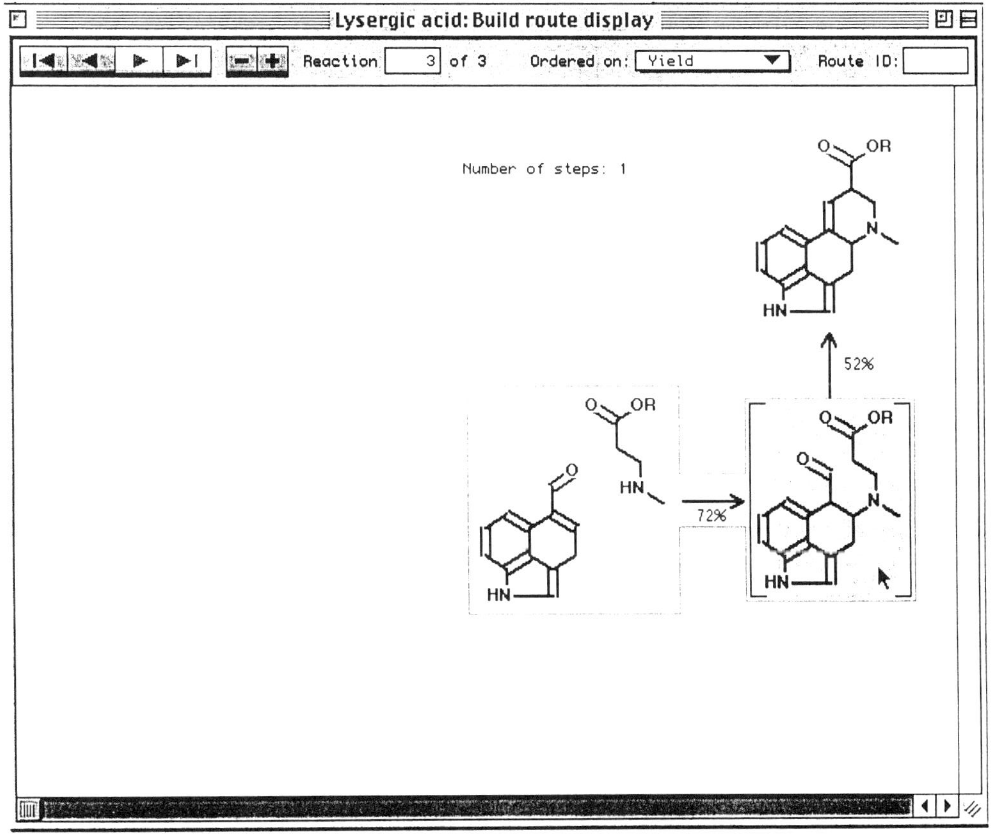

Figure 6. Build route reaction for last intermediate.

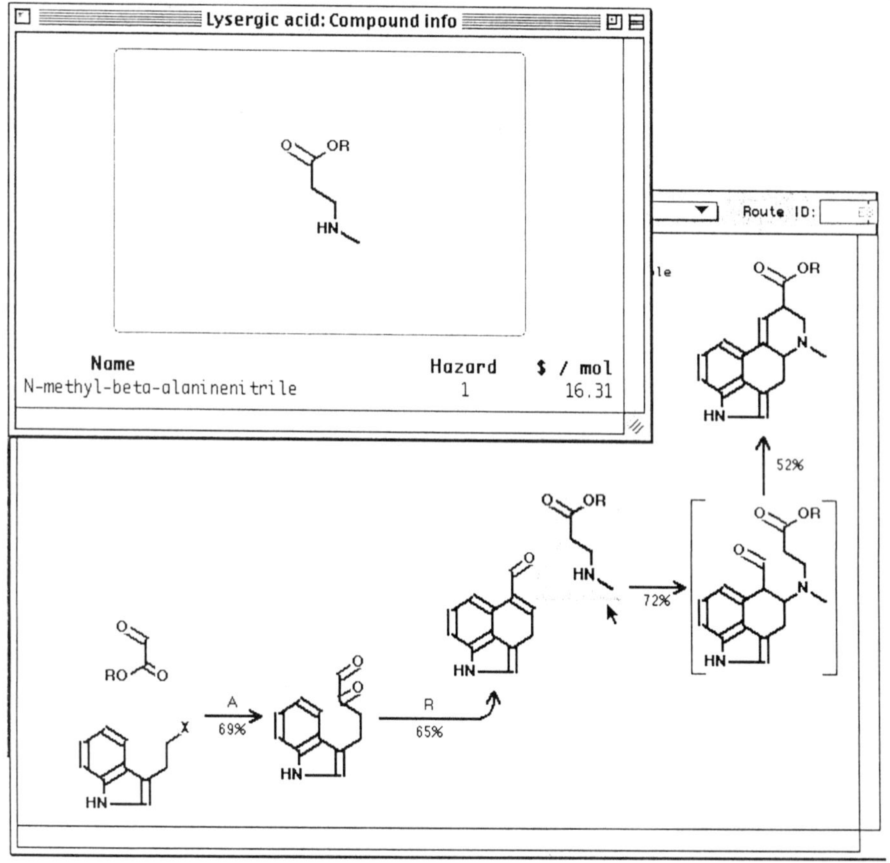

Figure 7. Optimal full route with starting material window.

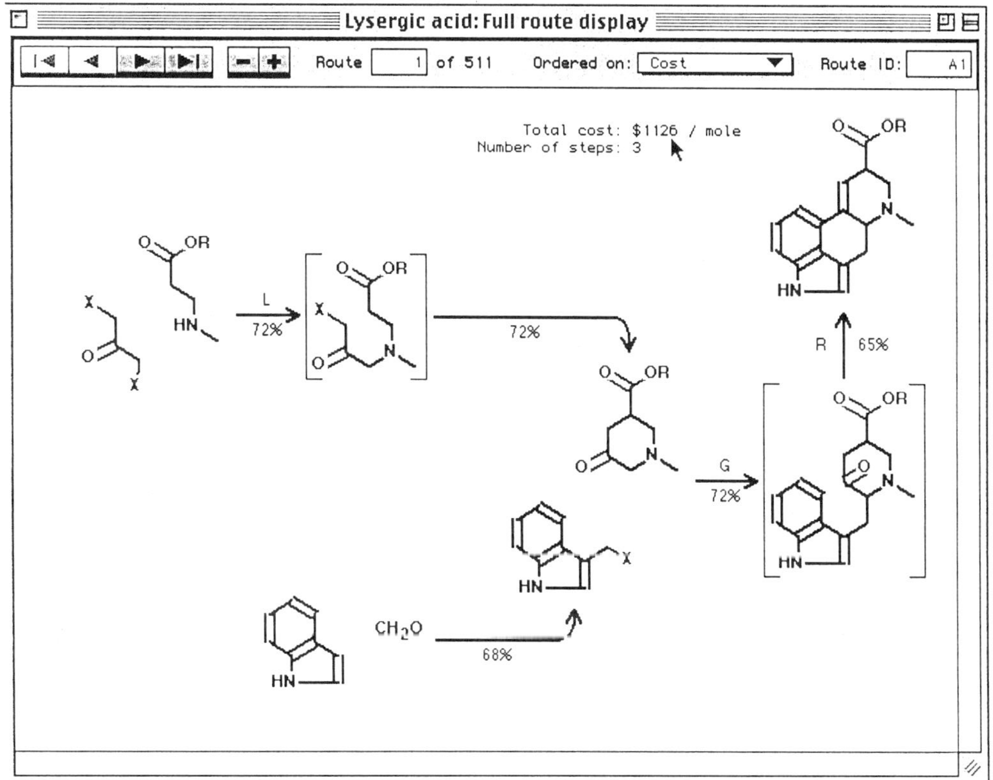

Figure 8. Optimal full route after deletions of intermediates.

Conclusions and Caution

Finally, it is apparent that inspection of many of the reactions generated will suggest that they are not viable when considered in detail. This should not be regarded as a shortcoming of the program, because SynGen does generate everything within the outlined criteria and the generalized format.

The power of SynGen is that it does explore all variants of the shortest, convergent syntheses from real starting materials. It creates all the themes of assembling the target from these, and if the chemistry in some route is not acceptable, it can frequently be adapted by adding protective refunctionalization or by making small changes in the molecules displayed.

SynGen is a program intended to open up new synthesis ideas that may not have been previously considered. When the details offered in the displayed routes are not considered to be workable, they can usually be repaired. The primary value of SynGen is to offer new perceptions of the variety of possible syntheses for any desired target molecule.

Acknowledgments

I wish to acknowledge long support by the National Science Foundation and most recently by the U.S. Environmental Protection Agency (Grant R825329). I am especially thankful to Dr. Camden Parks, who alone wrote the present SynGen program with great care and insight.

References

1. Hendrickson, J. B. *Angew. Chem. Intl. Ed.* **1990,** *29*, 1286-1295; *Anal. Chim. Acta* **1990,** *235*, 103-113; *Know. Eng. Rev.* **1997,** *12* (4), 1–18.
2. Hendrickson, J. B.; Sander, T. L. *J. Chem. Inf. & Comp. Sci.* **1995,** *35*, 251–260; *Eur. J. Chem.* **1995**, *1*, 449–453.
3. The WebReactions reaction retrieval program is used interactively from the web at http://www.WebReactions.net/, and it is a modification of the original search program named COGNOS *(2)*.
4. Ninomiya, I.; Kiguchi, T. In *The Alkaloids: Chemistry and Pharmacology*; Brossi, A., Ed.; Academic Press: New York, 1990; *38*, pp 1–164.

Chapter 11

Development of a Flexible System for the Simultaneous Conversion of Biomass to Industrial Chemicals and the Production of Industrial Biocatalysts

Johnway Gao, Brian S. Hooker, Rodney S. Skeen, and Daniel B. Anderson

Environmental Technology Division, Pacific Northwest National Laboratory, 902 Battelle Boulevard, P.O. Box 999, MS K2–10, Richland, WA 99352

A flexible system was developed for the simultaneous conversion of biomass to industrial chemicals and the production of industrial biocatalysts. In particular, the expression of a bacterial enzyme, beta-glucuronidase (GUS), was investigated using a genetically modified starch-degrading *Saccharomyces* strain in suspension cultures in starch media. Different sources of starch, including corn and waste potato starch, were used for yeast biomass accumulation and GUS expression studies under controls of inducible and constitutive promoters. A thermostable bacterial cellulase (E1 beta-1,4-endoglucanase) gene of *Acidothermus cellulolyticus* was also cloned into an episomal plasmid expression vector and expressed in the starch-degrading *Saccharomyces* strain.

Introduction

Tremendous amounts of biomass are produced worldwide by the agricultural and food-processing industries, with only a small portion directly used for food, fiber, chemicals, or energy. A large fraction of the remaining biomass provides only marginal economic benefit, and much of it must be managed as waste, incurring expenses to the producer or processor. For example, U.S. fruit and vegetable processors currently produce approximately 7.4 million dry metric tons of low-value byproduct (much of it sold at or near cost as livestock feed), and treat more than 300 million cubic meters of high biological oxygen demand (BOD) waste waters each year. Both of these byproduct streams are usually rich in starch, which could serve as an inexpensive yet robust feedstock for biotechnological processes for useful industrial enzyme production.

Natural yeast strains have been identified that can use starch as a primary growth substrate via complete or partial enzymatic hydrolysis (*1*, *2*). These yeast strains include *Saccharomycopsis fibuligera, Schwanniomyces castellii*, and *Saccharomyces diastaticus* (*1–3*). A fusion yeast cell strain of *Saccharomyces diastaticus* and *Saccharomyces cerevisiae* has also been demonstrated to degrade 60% of the starch present in culture media within two days (*4*). In addition, other natural *Saccharomyces* species can ferment starch and dextrin to ethanol (*5*). For the past two decades, this starch-biotransforming yeast technology was mostly used for producing single cell protein as a cattle feed supplement (*2*, *3*). Others use this technology to improve ethanol production from starch and higher sugars (*6*, *7*). In the current industries, ethanol can be produced as "green fuel" through a two-step biomass fermentation process, where the biomass is first hydrolyzed to a fermentable sugar, glucose, and the sugar is subsequently fermented to ethanol. A one-step fermentation process has been reported to use co-culture of a strong starch-degrading yeast strain, *Saccharomycopsis fibuligera*, and a robust ethanol producer, *Saccharomyces cerevisiae* for direct starch conversion to ethanol (*6*). The yeast fusion strain (*4*) is an alternative method for robust ethanol production directly from starchy biomass. The yeast strain *Saccharomyces cerevisiae* is unable to use starch during growth, while *Saccharomyces diastaticus* was able to use starch, and therefore, the fusion of these two yeast strains can provide simultaneous starch use and ethanol production.

Another one-organism system is also reported for direct starch hydrolysis and ethanol production using the strong starch degrading and moderate ethanol-producing yeast strain *Schwanniomyces castellii* (*7–9*). Moreover, the one-step fermentation process potentially can be achieved by cloning and expressing amylolytic enzyme genes into the robust ethanol-producer *Saccharomyces cerevisiae* (*10–12*). However, there is a lack of information on the production of recombinant industrial biocatalysts, excluding α-amylase and glucoamylase, directly from "waste" starch (biomass). Genetic modification of starch-degrading yeast strains can potentially provide a flexible system for simultaneous production of recombinant biocatalysts as intracellular products and chemical products as extracellular products directly from biomass materials. In addition to the intracellular recombinant biocatalysts and extracellular products, the remaining cell biomass (cell debris) eventually can be concentrated, dried, and sold as protein rich feedstock for cattle and poultry. The ideal design of such a process can become a "green process", because it will convert all the biomass to useful products and will eliminate process waste.

In the present study, a flexible system was proposed for the simultaneous conversion of biomass to industrial chemical products and the production of industrial biocatalysts using a starch-degrading yeast strain. In particular, a starch-degrading yeast strain was genetically modified for the expression of a bacterial enzyme, beta-GUS under the control of both inducible and constitutive promoters. In addition, the expression of a thermostable enzyme, endoglucanase, was also investigated in the starch-degrading yeast strain to demonstrate that this system may serve as means to produce inexpensive transgenic protein products.

Materials and Methods

Bacterial and Yeast Strains

An *Escherichia coli* competent cell strain, Top10 (Invitrogen Inc., Carlsbad, CA) was used as a host for routine cloning experiments. The yeast strain used in the study was a *Saccharomyces* strain, which is a yeast hybrid strain of *Saccharomyces cerevesiae* and *Saccharomyces diastaticus*, provided by Dr. James R. Mattoon of the University of Colorado in Colorado Springs, Colorado.

Cloning and Transformation Techniques

Plasmid DNA purification, plasmid construction and transformation, and polymerase chain reaction (PCR) were conducted under standard molecular cloning techniques (*13*). An EasySelect Expression Kit (Invitrogen, Carlsbad, CA) was used for preparing competent yeast cells, which were subsequently used for expression vector transformation. A total amount of 3 μg (in 5 to 10 μL) of expression vector DNA was used for each transformation of 100 μL competent cells. Upon transformation, the transformed yeast cells were plated onto a selective agar plate containing yeast extract, peptone, and dextrose (YPD). The selective YPD media consists of 1.0% dextrose (glucose), 0.5% yeast extract, and 1.0% peptone, and it is supplemented with 200 mg/L of the antibiotic Zeocin (Invitrogen). After a three-day incubation period at 30 °C, transformed yeast colonies were obtained on the selective agar plate.

Culture and Expression Medium

For protein expression under inducible conditions, the recombinant yeast strain was first grown in the growth medium for cell biomass accumulation. The growth medium contains waste potato starch (Lamb-Weston Inc., Richland, WA) or corn starch (Sigma, St. Louis, MO) as the primary carbon source, supplemented with 1.0% peptone and 0.5% yeast extract as the nitrogen source. The expression medium contains 2% galactose, 1.0% peptone, and 0.5% yeast extract. The medium was adjusted to pH 6.0 by hydrochloric acid and autoclaved at 250 °F for 20 minutes. For protein expression under constitutive conditions, the expression or production medium is the same as the growth medium without the inducing agent, galactose. All suspension cultures were grown aerobically at 30 °C in an orbital shaker shaking at 200 rpm.

Protein Sample Extraction and Foreign Protein Analysis

Intracellular protein of transformed yeast biomass was extracted using the glass-bead disintegrating method. Briefly, in this method one volume of acid-washed 500 μm glass beads (Sigma) was added to one volume of cell sample in an extraction buffer containing 50 mM pH 7.0 sodium phosphate, 1 mM ethylenediaminetetraacetic acid (EDTA), 1 mM phenylmethylsulfonyl fluoride (PMSF), 10 mM beta-mercaptoethanol, and 0.1% triton X-100. Cells were disrupted by vortexing vigorously for 30 seconds for five times. Samples were kept on ice during vortexing intervals. After cell disruption, the sample was centrifuged at 4 °C at 20,000*g* for 5 minutes. The supernatant was saved for both protein and enzyme activity assays. The extracted protein samples were assayed for protein concentration using the Bio-Rad protein assay (Bio-Rad Laboratories, Hercules, CA).

Recombinant beta-glucuronidase activity was assayed using an enzymatic reaction in which a substrate 4-methylumbelliferyl-beta-D-glucuronide (MUG) can be hydrolyzed by glucuronidase to form a fluorescent compound, 4-methylumbelliferone (MU). One unit of glucuronidase activity is defined as the amount of glucuronidase that produces one pmol MU from MUG per minute at 37 °C. The fluorescence of produced MU was assayed in a DyNA QUANT 200 fluorometer (Pharmacia Biotech, Piscataway, NJ). The specific activity of glucuronidase is calculated as the units of glucuronidase per milligram of total protein in the sample.

Recombinant E1 Endoglucanase activity was assayed using an enzymatic reaction in which the substrate 4-methylumbelliferyl-beta-D-cellobioside (MUC) can be hydrolyzed by E1 endoglucanase to form a fluorescent compound, 4-methylumbelliferone (MU). One unit of E1 endoglucanase activity is defined as the amount of E1 endoglucanase that produces one pmol MU from MUC per minute at 55 °C. The specific activity of E1 endoglucanase is calculated as the units of E1 endoglucanase per milligram of total protein in the sample.

Results and Discussions

Plasmid Vector Construction for Beta-Glucuronidase Expression

Plasmid vectors were constructed to effectively transform the starch-degrading *Saccharomyces* strain and to select the transformants after transformation. The Plasmid construction flowchart is shown in Figure 1. Two plasmid expression vectors were constructed respectively for inducible and constitutive GUS expression as shown in Figure 2 (A) and (B). The inducible expression vector, pGA2026, was constructed by replacing the ampicillin resistance and URA3 genes in a plasmid vector pYES (Invitrogen, Carlsbad, CA) with the antibiotic Zeocin resistance gene of the plasmid vector pGAPZα-A (Invitrogen). The plasmid pGA2026 enables the selection of transformed *Saccharomyces* strains without using a uracil-deficient selection medium. The plasmid pGA2026 contains an expression cassette for foreign gene expression under the control of an inducible galactokinase promoter (GAL1) P_{GAL1} and a T_{cyc1} terminator. In addition, the plasmid also contains a 2 µm DNA fragment for plasmid replication in *Saccharomyces* strains, a ColE1 origin for plasmid replication during gene manipulation in *E. coli* strains, an f1 phage origin, and the antibiotic Zeocin resistance gene for both transformed yeast and *E. coli* selection during gene manipulation and after transformation. A bacterial glucuronidase gene (*gus*) (*14*) was subsequently cloned into pGA2026 at Hind

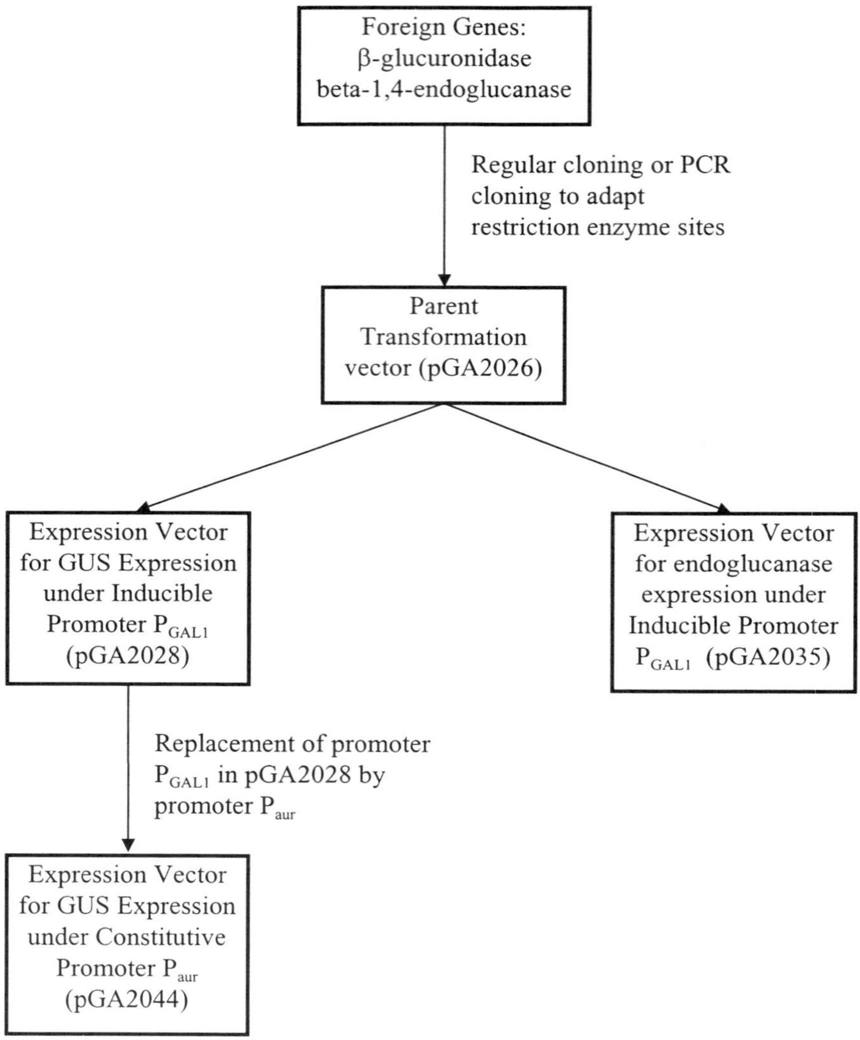

Figure 1. Flowchart for plasmid vector constructions.

III (H) and Sac I (H) restriction enzyme sites to form the GUS expression vector pGA2028.

For the constitutive GUS expression in the yeast host, a constitutive promoter P_{aur} was used. The gene blast analysis shows that the P_{aur} promoter locates in *Saccharomyces cerevesiae* chromosome XI (*15*) and controls the expression of the antibiotic resistance gene aureobasicin A (*16*). The P_{aur} promoter was cloned from a vector pAUR101 (PanVera Corp., Madison, WI) by PCR using the following primers: 5' end forward primer-GTC GAC GGT ACC CAG CTG TGT AAA GTT CTT CTT CCA GAT TGT GCG CAA ACT and 3' end reverse primer-GAT TAA AAG CTT TTT AAA ATA TGA AAA CCG CAA. One restriction enzyme site Pvu II was adapted at the 5' end of the promoter. The PCR'd promoter was disgested with the restriction enzymes Hind III (H) and Pvu II (P) and was subsequently cloned into the expression vector pGA2028 at the Hind III and Pvu II sites to replace the GAL1 promoter to form a constitutive GUS expression vector pGA2044, as shown in Figure 2 (C).

Plasmid Vector Construction for Beta-1,4-endoglucanase Expression

The starch-degrading yeast strain *Saccharomyce* was also used to express a thermostable cellulolytic enzyme gene, endoglucanase from *Acidothermus cellulolyticus* (*17*). The E1 beta-1,4-endoglucanase precursor gene was obtained from Steven R. Thomas of the National Renewable Energy Laboratory in Golden, Colorado. The mature endoglucanase gene was cloned out by PCR from the beta-1,4-endoglucanase precursor gene and was adapted with a Hind III restriction enzyme site and an initiation codon ATG at the 5' end and a Sac I restriction enzyme site at the 3' end using the following primers: 5' end forward primer-AGG CCT AAG CTT ATG GCG GGC GGC GGC TAT TGG CAC ACG, and 3' end reverse primer-GTC GAC GAG CTC TTA ACT TGC TGC GCA GGC GAC TGT CGG. The PCR'd mature E1 gene was digested with Hind III (H) and Sac I (S) restriction enzymes and cloned into the plasmid vector pGA2026 at Hind III (H) and Sac I (S) restriction enzyme sites to form pGA2035, as shown in Figure 2 (D).

Beta-glucuronidase Expression in Hexose Culture Medium

To test glucuronidase gene expression after the pGA2028 vector transformation, ten transformed-yeast colonies were streak-purified on fresh selective YPD agar plates and single colonies were used in batch cultures for glucuronidase expression. The yeast colonies were first grown aerobically in 2

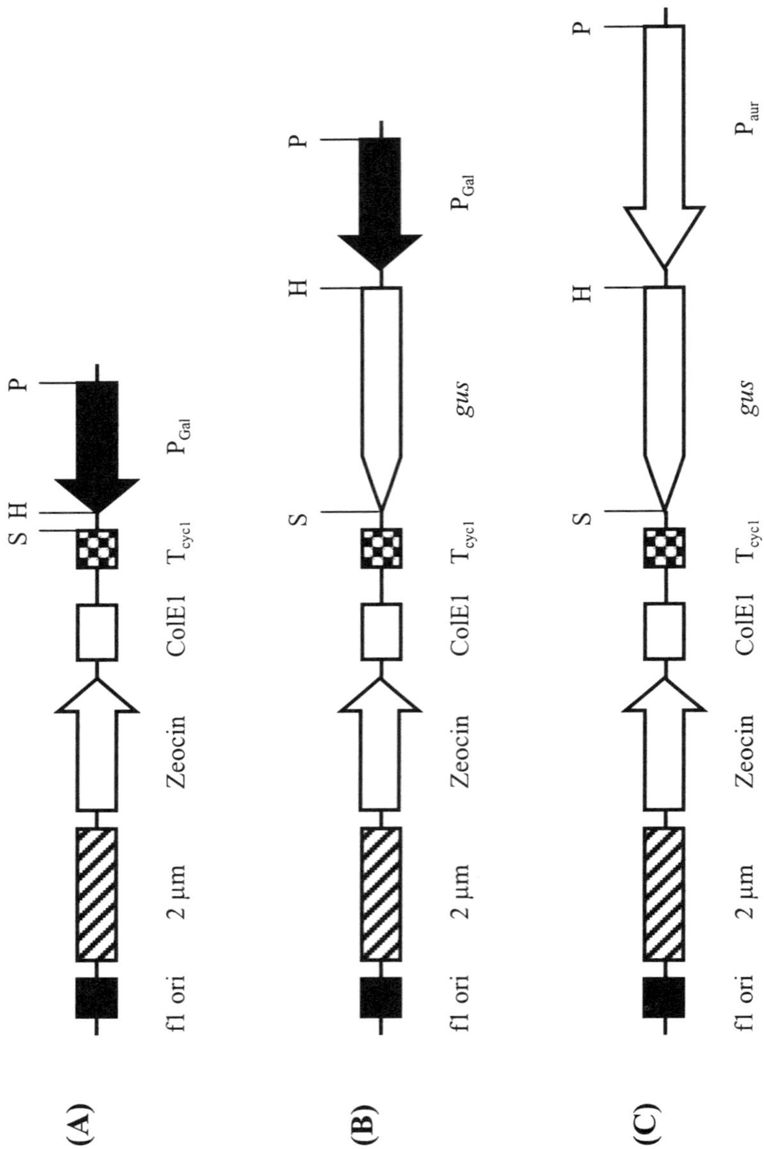
(A)
f1 ori
2 µm
Zeocin
ColE1
Tcyc1
S H
PGal
P
(B)
f1 ori
2 µm
Zeocin
ColE1
Tcyc1
S
gus
H
PGal
P
(C)
f1 ori
2 µm
Zeocin
ColE1
Tcyc1
S
gus
H
Paur
P

(D)

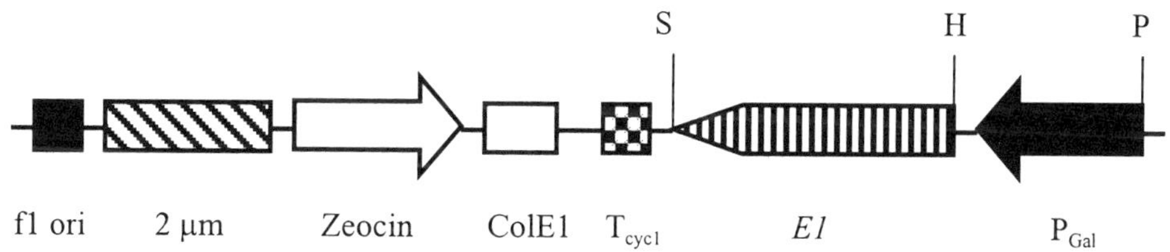

Figure 2. Plasmid vectors for the transformation of starch-degrading Saccharomyces strain and the expression of beta-glucuronidase and E1 endoglucanase. (A): plasmid vector pGA2026 for Saccharomyces transformation; (B): plasmid vector 2028 for glucuronidase expression under the control of P_{Gal} promoter; (C): plasmid vector pGA2044 for glucuronidase expression under the control of P_{aur} promoter; (D) Plasmid vector pGA2035 for the expression of E1 endoglucanase under the control of P_{Gal} promoter.

mL of YPD medium for 16 hours. The propagated biomass was transferred into a 2 mL production medium containing promoter activity inducer, galactose for GUS expression. After a 5-hour inducing period, yeast biomass was harvested, and intracellular protein was extracted using the glass-bead disintegrating method. Table I illustrates the results of GUS specific activities for ten different transformed clones. These data demonstrate that glucuronidase can be highly expressed in the transgenic yeast host under the control of GAL1 promoter. The highest specific activity obtained in the culture was 10,057 units per milligram of extracted intracellular protein. There was no extracellular GUS activity found since there was no secretion transit peptide sequence to the *gus* gene.

Table I. Recombinant GUS Activity in Transgenic Starch-Degrading *Saccharomyces* Strain

Clone Number	*GUS Specific Activity (units/mg)*
Control	16
1	3469
2	10,057
3	4061
4	7013
5	5309
6	7786
7	6035
8	4439
9	4000
10	8346

NOTE: Control sample has no genetic transformation

Beta-Glucuronidase Expression in Starch Culture Medium

A starch medium was used to cultivate transformed yeast clones for GUS expression. Corn and waste potato starches were used. The culture medium is sugar-free and contains 1.0% corn or potato starch supplemented with 1.0% peptone and 0.5% yeast extract. Clone number 2 in Table I was used in the GUS expression test. After a two-day growth period in the potato starch medium, the biomass was collected, and glucuronidase expression was induced in a production medium primarily containing 2% galactose as an inducer. Cells were harvested periodically, and intracellular protein was extracted in the extraction buffer. Figure 3 shows the results of GUS expression within different inducing periods. The glucuronidase activity reached its highest expression after a four-hour induction period and leveled off thereafter, indicating stable expression of GUS in the culture. In addition to potato starch, corn starch was also used to propagate cell biomass, and the glucuronidase activity was subsequently induced in the galactose medium. Table II shows the results of glucuronidase activities after a four-hour induction in the galactose medium, using glucose medium as the control. In both cases, the GUS specific activities were 84% and 75% of that obtained in the glucose culture medium, respectively, for potato and corn starch media. The advantage of using inducible promoter is to accumulate the cell biomass first and subsequently to induce the production of desired products. This is especially useful when the production process tries to avoid the accumulation of certain products that may inhibit cell growth. After the biomass reaches the required level, the addition of an inducer will initiate the accumulation of large amounts of final product without interruption of cell biomass accumulation, which is essential to the product productivity. However, the addition of an inducer sometimes complicates the production process and adds additional cost to the final product.

Table II. GUS Induction in Transformed Yeast Cells after Growth in Different Media

Growth Medium	GUS Specific Activities (units/mg)
Glucose	15,344 ± 26
Potato starch	12,933 ± 697
Corn starch	11,478 ± 3345

NOTE: Results were averaged from duplicate experiments, and "±" stands for standard error.

Beta-Glucuronidase Expression Under Constitutive Condition

The expression of beta-glucuronidase under the control of promoter P_{aur} was tested in both glucose and starch media without the inducing substrate, galactose. After being transformed with plasmid expression vector pGA2044, transformed yeast colonies were picked and streak-purified on fresh selective YPD agar plates containing 200 mg/L antibiotics Zeocin. Single colonies were used in batch cultures for the glucuronidase expression test. Table III shows the expression results of beta-glucuronidase in glucose growth medium in the absence of galactose after 20-hour growth. In this test, the yeast clone transformed with plasmid expression vector pGA2028 with the inducible promoter P_{Gal} served as the control. Experimental results indicated that recombinant specific glucuronidase activities under the constitutive promoter P_{aur} were about 7 to 15 times higher than that under the control of the inducible promoter P_{Gal} in the test condition without an inducing substrate. The use of inducing substrate, in general, may add additional cost to an enzyme production process. The application of constitutive promoter for foreign protein expression eliminates the use of inducing agent and therefore simplifies the production process.

Figure 4 shows a batch culture process of glucuronidase expression under the constitutive promoter P_{aur} was investigated in a 250-mL Erlenmeyer flask with 50 mL of starch culture medium. The medium contains 1.0% waste potato starch, 0.5% yeast extract, and 1.0% peptone. The culture was inoculated with 1.0% overnight seed culture grown in the same starch medium. Cell growth and glucuronidase activities were monitored periodically. The results showed that glucuronidase specific activity was proportional to the cell growth in the culture, indicating that the glucuronidase expression was growth associated during the testing period. The culture reached a specific glucuronidase activity of about 4200 units/mg at the end of the test.

Table III. GUS Expression in Transformed Yeast Cells in Glucose Media

Test Number	*Plasmid No.*	*Promoter*	*GUS Specific Activities (units/mg)*
1	pGA2028	P_{Gal}	486
2	pGA2044	P_{aur}	4839
3	pGA2044	P_{aur}	4044
4	pGA2044	P_{aur}	7931

Table IV shows the results of glucuronidase expression under P_{aur} promoter in both glucose and starch medium. The cultures were conducted in 250-mL Erlenmeyer flasks with 50 mL of starch culture medium, containing 1% waste potato starch, 0.5% yeast extract, and 1.0% peptone. Results showed that both glucose and starch media were favorable for glucuronidase expression, while starch medium induced a 35% higher glucuronidase expression than in the glucose medium.

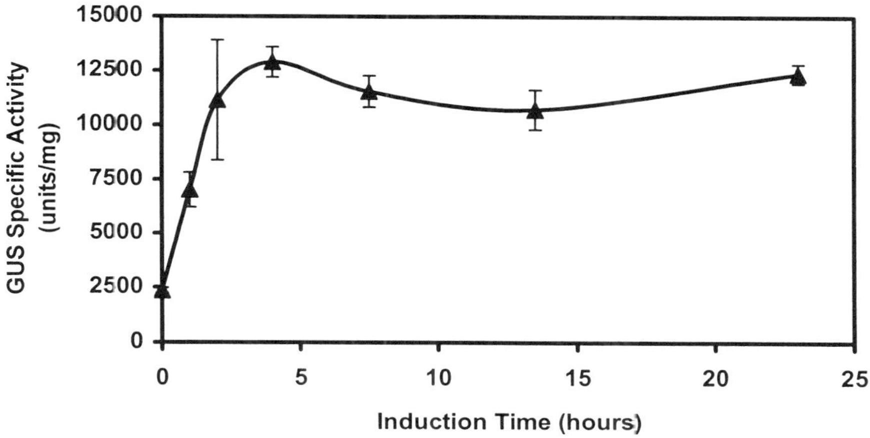

Figure 3. Glucuronidase induction test in transformed starch-degrading Saccharomyces strain using the production medium after cells were propagated in the growth medium. Each data point was averaged from duplicate tests. Error bars represent standard error.

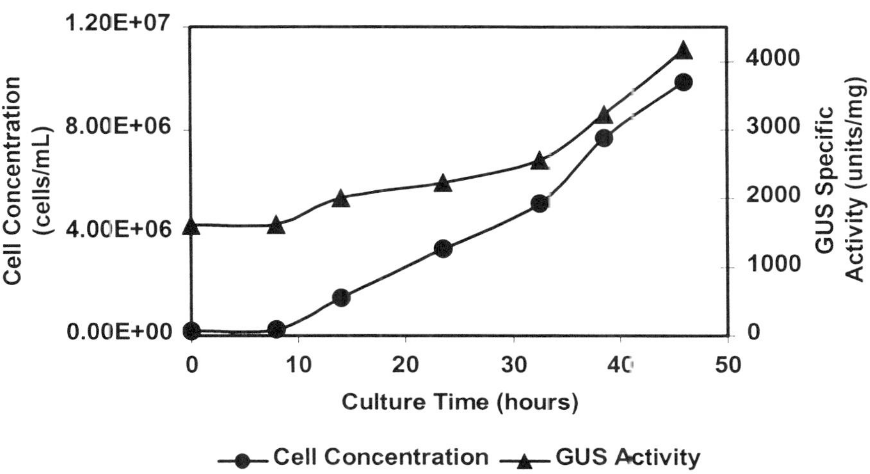

Figure 4. Glucuronidase expression under the constitutive promoter P_{aur} in a batch culture of transformed starch-degrading Saccharomyces strain. The medium contains 1% waste potato starch, 0.5% yeast extract, and 1% peptone. Each data point was averaged from duplicate tests.

Table IV. GUS Expression in Yeast Cells Transformed with Plasmid pGA2044

Culture Medium	*GUS Specific Activities (units/mg)*
Glucose	4863 ± 111
Potato Starch	6608 ± 1282

NOTE: Results were averaged from duplicate experiments, and "±" stands for standard error.

Beta-1,4-endoglucanase Expression in Hexose Culture Medium

The starch-degrading yeast strain *Saccharomyces* was also used for pGA2035 vector transformation and E1 endoglucanase expression. Upon transformation, the transformed yeast cells were plated onto selective YPD agar media containing glucose, yeast extract, peptone, and the antibiotic Zeocin. After a 3-day incubation period at 30 °C, transformed yeast colonies were obtained in the selective culture media.

Twelve transformed yeast colonies were streak-purified on fresh selective YPD agar plates, and single colonies were used in batch cultures for E1 endoglucanase expression. The yeast colonies were first grown aerobically in a 2-mL YPD medium for 16 hours. The propagated biomass was transferred into a 2-mL production medium containing the promoter activity inducer, galactose for E1 endoglucanase expression. After a 5-hour inducing period, yeast biomass was harvested and intracellular protein was extracted in the extraction buffer using the glass-bead disintegrating method. Table V shows the results of E1 endoglucanase specific activities of twelve different transformed clones. The highest specific activity obtained in the culture was 1724 units per milligram of extracted intracellular protein. However, there was no secreted E1 endoglucanase detected in the culture medium even though a native leader sequence was used in the E1 expression construct. This is probably due to the strong binding activity of the E1 cellulose-binding domain to the cell wall cellulose, inhibiting E1 secretion. Preliminary results showed that the removal of the cellulose-binding domain greatly enhanced E1 enzyme secretion.

Conclusions

The results presented in this research demonstrate that starch or waste starch can be used as an inexpensive media for the production of recombinant

Table V. Recombinant E1 Endoglucanase Activity in Transgenic Starch-Degrading *Saccharomyces* Strain

Clone Number	*E1 endoglucanase Specific Activity (units/mg)*
Control	7
1	1724
2	961
3	1241
4	1383
5	1294
6	1273
7	1257
8	1111
9	1501
10	1258
11	508
12	1476

NOTE: Control sample has no genetic transformation.

biocatalysts using a starch-degrading yeast strain. With genetic manipulation methods, the starch degrading yeast strain can express foreign protein products. The expression of a bacterial enzyme glucuronidase could reach a specific activity of 12,903 units per milligram of extracted intracellular protein under an induced condition after the transformed yeast strain grew in a starch medium. Another enzyme, E1 endoglucanase, was also expressed in the starch-degrading yeast strain at the level of 1724 units per milligram of extracted protein.

The application of the constitutive promoter P_{aur} has the advantage of eliminating the addition of an inducing agent such as galactose in the production medium, therefore simplifying the production process and reducing the production cost. As compared to the glucose medium, the recombinant protein expression level was higher than that in the starch medium when the constitutive promoter P_{aur} was used.

The "waste" biomass, such as byproduct starch materials from potato processors and wheat and corn mills, can serve as a primary carbon source for the production of useful and useable industrial chemicals, including biocatalysts. Such a process has three advantages: (1) use of inexpensive biomass as a feedstock for the production process; (2) use of a one-step process without the need for an additional hydrolysis process (such as an acid hydrolysis process) for starchy materials; and (3) elimination of process waste. With advanced genetic engineering technologies and the use of an inducible or constitutive flexible system, the production of useful industrial products becomes possible and can provide a manufacturer with more opportunities to produce varied products under different production conditions.

Acknowledgments

The authors thank Dr. James R. Mattoon of the University of Colorado at Colorado Springs, CO for providing the *Saccharomyces* strain. The authors also thank Dr. Steven R. Thomas of National Renewable Energy Laboratory, Golden, CO for providing the gene of a thermostable cellulolytic enzyme, endoglucanase from *Acidothermus cellulolyticus*. This work was supported by the U.S. Department of Energy, Office of Science and Technology, under Contract DE-AC06-76RLO 1830. The Pacific Northwest National Laboratory is operated by Battelle for the United States Department of Energy under contract DE-AC06-76RLO 1830.

References

1. Sills, M.; Stewart, G. G. *J. Inst. Brew.* **1982,** *88*, 313.
2. Hongpattarakere, T.; H-Kittikun, A. *J. Microbiol. Biotechnol.* **1995**, *11,* 607.
3. Lemmel, S. A.; Heimsch, R. C.; Korus, R. A. *Appl. Environ. Microbiol.* **1980,** *39,* 387.
4. Kim, K.; Park, C. S.; Mattoon, J. R. *Appl. Environ. Microbiol.* **1988,** *54*, 966.
5. Laluce, C.; Bertolini, M. C.; Ernandes, J. R.; Martini, A. V.; Martini, A. *Appl. Environ. Microbiol.* **1988,** *54*, 2447.
6. Pirselova, K.; Smogrovicova, D.; Balaz, S. *World J. Microbiol. Biotechnol.* **1993,** *9*, 338.
7. Ryu, Y. W.; Ko, S. H.; Byun, S. Y.; Kim, C. *Biotechnol. Lett.* **1994**, *16,* 107.
8. Saucedo-Castaneda, G.; Lonsane, B. K.; Navarro, J. M.; Roussos, S.; Raimbault, M. *Appl. Biochem. Biotechnol.* **1992,** *36,* 47.
9. Saucedo-Castaneda, G.; Lonsane, B. K.; Krishnaiah, M. M.; Navarro, J. M.; Roussos, S.; Raimbault, M. *Process Biochem.* **1992,** *27,* 97.

10. Janse, B. J. H; Pretorius, I. S. *Appl. Microbiol. Biotechnol.* **1995**, *42*, 878.
11. de Moraes, L. M. P.; Astolfi-filho, S.; Oliver, S. G. *Appl. Microbiol. Biotechnol.* **1995**, *43*, 1067.
12. Rothstein, S. J.; Lazarus, C. M.; Smith, W. E.; Baulcombe, D. C.; Catenby A. A. *Nature* **1984,** *308*, 662.
13. Sambrook, J., Fritsch, E. F., Maniatis, T., Eds. *Molecular Cloning: A Laboratory Manual*, 2nd ed.; Cold Spring Harbor Laboratory Press: Plainview, NY, 1989.
14. Jefferson, R. A.; Kavanagh, T. A.; Bevan, M. W. *EMBO J.* **1987**, *6*, 3901.
15. Dujon, B.; Alexandraki, D.; Andre, B.; Ansorge, W.; Baladron, V.; et al. *Nature* **1994**, *369,* 6479.
16. Hashida-Okado, T.; Ogawa, A.; Kato, I.; Takesako, K. *FEBS Lett.* **1998**, *425* (1), 117.
17. Laymon, R. A.; Himmel, M. E.; Thomas, S. R. **1995**, *GeneBank* Database, NID 988299.

Chapter 12

Radiation Chemistry: The Basis for an Inherently Green Process Technology

Anthony J. Berejka

Ionicorp[+], 4 Watch Way, Huntington, NY 11743 (email: berejka@msn.com)

In using ionizing radiation generated from electron beam accelerators or in using photoinitiated reactions started by high intensity ultraviolet light, industry has eliminated the need for volatile organic compounds from coatings, inks, and adhesives. Such processes have also been found to be energy efficient and to minimize wastes, conserving both energy and materials. Large-scale process studies have indicated that ionizing radiation can also be used to precipitate sulfur dioxide and nitrous oxides from stack gases and to dehydrohalogenate toxic halocarbons found in wastewater and in soils. While recognized for its environmental benefits by organizations such as the International Atomic Energy Agency, the United States Environmental Protection Agency, and the National Science Foundation, industrial acceptance of such inherently green processes based on radiation chemistry has been slowed by a tendency in U.S. industry to conserve capital in plants and equipment rather than to make fundamental changes in process technology. Case studies are presented to illustrate the mechanisms involved and the resultant environmental and economic benefits.

Introduction

Radiation chemistry embraces a broad scope of scientific and technical inquiry. However, except for the selected use of gamma emission from ^{60}Co for medical device sterilization and for its emerging use for food treatment, industry does not use radioactive sources. Radioactive ^{60}Co, with its 5.3-year half-life, poses concerns over transport, storage, and waste disposal, as well as a continuing loss of activity, even when not in use for its intended purposes. Thus, industrial radiation processing relies primarily on two electrically generated radiation sources: accelerated electrons and photons from high intensity ultraviolet lamps. Accelerated electrons have the ability to penetrate matter, being stopped only by mass. In contrast, high intensity ultraviolet light generates primarily surface effects, as illustrated in Figure 1 (*1*).

As shown in Table I, there are over 1000 high-current electron beam accelerators now in industrial use, ranging in voltage potential from 70 keV up to 10 MeV. With electron penetration being proportional to acceleration voltage, as shown in Figure 2, the lower voltage accelerators find use in surface curing and in the cross-linking of polymeric films. Mid-voltage accelerators are used to cross-link the jacketing on wire and cable, in the manufacture of heat shrink tubing, and to partially cross-link tire components. High voltage 10 MeV accelerators are being used in the cross-linking of matrix materials in carbon fiber composites and are being evaluated for food processing applications and medical device sterilization (*2, 3*).

Given their significantly lower capital costs, there are well in excess of 100,000 high intensity ultraviolet units now in industrial use. Their applications deal mainly with the surface curing of inks, coatings, and adhesives. These units typically operate in the 200 to 450 nm range, with lamp intensities as high as 240 W/cm of lamp length (*1*).

Table I. Industrial Use of Electron Beam Accelerators

Voltage Range	*Industrial Units*	*Average Power*
0.1 to 0.3 MeV	300	100 kW
0.5 to 5.0 MeV	700	40 kW
5.0 to 15 MeV	35	15 kW

A given application area dictates the preference for a specific accelerator voltage. The use of ionizing radiation to cross-link wire and cable jacketing and heat shrinkable tubing remains the dominant market (36%) for industrial

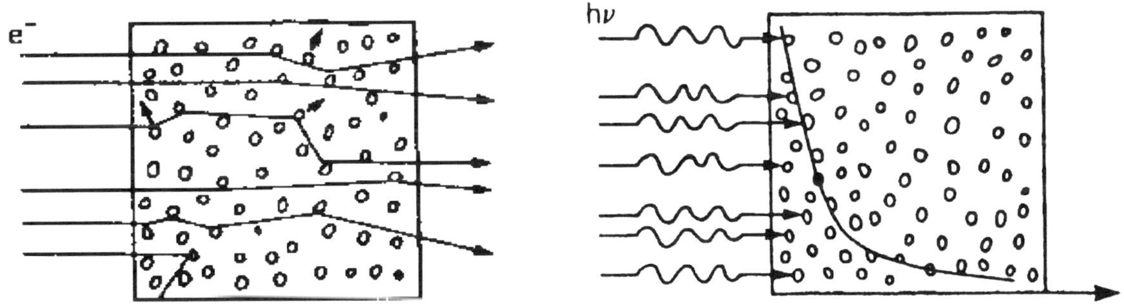

Figure 1. Electron beam and ultraviolet light penetration.

irradiation processing, as illustrated in Figure 3, with 1.5 MeV accelerators being widely used in the wire and cable industry. Heat shrinkable films used for food packaging rely upon lower voltages, such as 300 keV to 800 keV, and account for a significant market end use (23%). These applications, for the most part, depend upon the favorable cross-linking response of polyethylene to irradiation. Because ionizing radiation from an electron accelerator can be tightly controlled, major tire manufacturers use irradiation processing to partially cross-link components to enhance their toughness before being molded into a finished tire. This reduces cord strike-through and displacement during molding operations.

While some service centers are used by smaller wire and cable producers, a growing interest in irradiation processing through toll service centers is developing for medical device sterilization and, on the horizon, for food disinfection. These application areas demand high beam penetration and the higher voltage 10 MeV accelerators.

The fastest growing market segment (25%) has been that of low-voltage (70 to 300 keV) self-shielded accelerators for the curing of inks, coatings, and adhesives. Radiation chemistry is used in situ to polymerize liquid-applied materials containing near-zero volatile organic compounds (VOCs) into functionally cured materials and products.

Green Chemistry for Inks and Coatings

VOC Elimination Plus Energy Efficiency

The increased industrial acceptance of high-intensity ultraviolet light processing and the use of low voltage electron beams (EB) has been stimulated by the Clean Air Act Amendments of 1990. Historically, formulations for inks, coatings and adhesives have relied on the use of polymeric resins dissolved in volatile organic solvents to reduce the viscosity needed for application. Systems with comparable finished properties have been developed based on compositions of non-volatile monomers and oligomers, typically <10,000 in molecular weight. For the most part, the chemistries involved rely upon the vinyl termination of acrylated moieties. There are, for example, acrylate terminated urethane, polyester, and epoxy oligomers. Required application viscosities are achieved by diluting such oligomers with appropriate monomers or monomer combinations without the use of volatile organic solvents. As outlined in Table II, such liquid materials are converted to dried inks or coatings upon exposure to an electron beam or, when formulated with photoinitiator compounds, to ultraviolet light. These materials have all of the functional

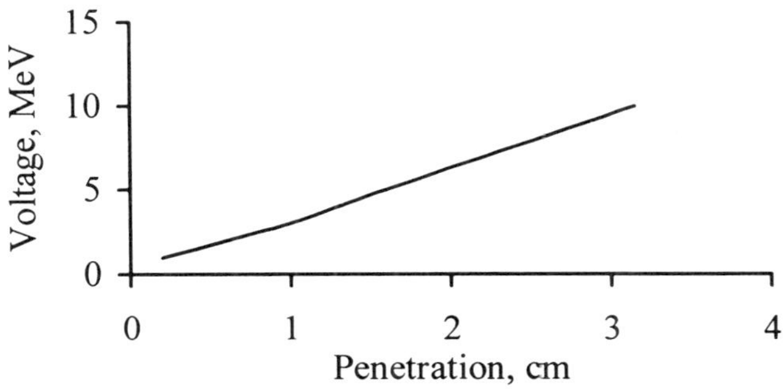

Figure 2. Electron beam penetration.

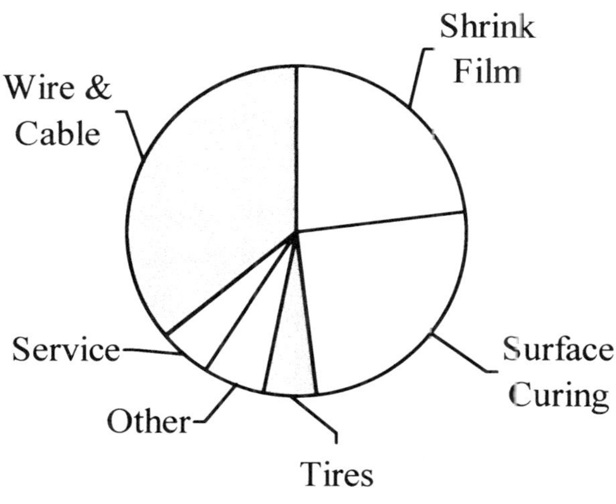

Figure 3. Industrial use of electron beam processing.

properties of the more traditionally used solvent-borne systems but without the dependence on VOCs. The dominant mechanisms involved are free-radical conversion of double bonds to form higher molecular weight or fully cross-linked polymeric materials (*1*).

Table II. Components of Industrial Coatings

Conventional	*UV/EB Coatings*
Binder	Binder
Pigment	Pigment
Solvent = VOCs	UV/EB Radiation
Heat	
Drying Time	

The merits of such systems have been noted by former U.S. EPA Administrator Carol Browner in her February 12, 1997 testimony before the United States Senate. In November 1998, the South Coast Air Quality Management District, which is responsible for air quality in the Los Angeles basin, has officially recognized UV/EB curing as an environmentally friendly process and has exempted such processes from some of its permitting requirements.

Wood Coatings

In some states, including New York, there has been a confluence of environmental concern through enforcement and economic support for environmentally friendly or "green" process technologies. The New York State Energy Research and Development Authority has funded several projects involving UV/EB processing. In one case, a small family-owned business in the Catskill Mountains producing a common wood product, brush blocks (i.e., the base of push brooms), received about $160,000 in financial assistance to obtain a spray coating/UV curing process line, shown in Figure 4. This firm was thus able to eliminate its use of 20%-solids nitrocellulose lacquers and to convert to UV processing. This not only eliminated VOCs as an air pollutant, but also reduced production concerns about solvent vapors that irritated workers, the need for segregated coating rooms and facilities, and the like. However, the main advantage in converting to UV curing in coating these small wood products was found in enhanced production efficiency. This particular process line is capable of instant start up and shutdown, thereby minimizing any process

Figure 4. Spray coating/UV curing process line.

waste. Because the UV cure is nearly instantaneous, production rates were increased over 6-fold. A full palate of coated wooden pieces, which would have taken over half a day to coat and dry using a solvent system and infrared heating, could be finished in less than 45 minutes. While the base cost for a UV-curable coating was considerably more than for a traditional solvent-borne lacquer, on an applied cost basis there was an actual savings on the cost per coated part, which included not only materials, but labor savings as well (*4*).

Energy Efficiency

In addition to state and regional utility-related organizations, the Electric Power Research Institute also has been supportive of the electrotechnolgies of UV/EB processing. Ultraviolet and electron beam processing represent a clean and efficient use of electrical power, as illustrated in Table III. In contrast to water-borne technologies, process systems based upon radiation chemistry represent a truly "green" technology. All water-based processes, whether used in converting industries to reduce dependence upon organic solvents and their emissions or even in synthesis schemes, are inherently energy inefficient, because ultimately, water must be evaporated and driven off to produce a dried product or material (*5*).

Table III. Energy Needed to Dry/Cure Coatings

Coating System	*Energy Required (cal/g)*
Water Based	810
Solvent Based	177
EB Curable	7

Tape and Label Manufacture

Comparable energy savings for electron beam and ultraviolet curing can also be found in the manufacture and coating of pressure sensitive adhesives for tapes and labels. Moderate-cost raw materials, some of which are common to solvent-borne and water-borne systems, such as hydrocarbon tackifiers, can be used in the development of radiation-curable pressure sensitive adhesives. Tapes coated with such and cured using ultraviolet light have demonstrated properties equal to commercial solvent-borne adhesive systems.

As presented in Table IV, cost analyses of the capital equipment used in the manufacture of the historic solvent-borne pressure sensitive adhesives and of hot-melt pressure sensitives that in themselves are also near-zero in VOCs have shown that radiation-curable adhesives require less investment in equipment to manufacture and consume less power during manufacture than these alternative systems. The compact space needed for either an electron beam or ultraviolet curing unit minimizes plant floor space requirements. An adhesive coating line with electron beam or ultraviolet curing capability can occupy one-ninth or less of the floor space required (for example 5 m versus 45 m in length) for similar coating installations that have long, energy-inefficient drying ovens. While some major producers of pressure sensitive tapes and labels have adopted radiation curing, there remains considerable growth potential for the use of this environmentally friendly process in these market areas (*6, 7*).

Table IV. Alternative Compounding Approaches for Adhesives

Objective: Produce at least 190 kg/h of adhesive for coating			
		Process	
	Solvent Based	*Hot Melt*	*UV/EB Curable*
Compounding equipment	Two-roll mill 55 cm x 1.5 m + 2800 L rubber dissolver	Continuous twin screw compounder (50 mm screws)	150 L change can mixer
Approximate capital costs	\$225,000 + \$90,000	\$310,000	\$70,000
Rated horsepower	150 hp + 30 hp sweeper; 75 hp cutter	40 hp	7.5 hp planetary; 10 hp disperser
Maximum power consumption/h	112 kWh + 78 kWh	30 kWh	13 kWh
Environmental constraints	Permitting for solvent use	None	None
Ancillary equipment	Storage drums	Automated weigh feeders and pump \$150,000	Extra vessel \$10,000 + liquid pumps \$26,000

Green Chemistry for Environmental Remediation

Wastewater Decontamination

One of the legacies of industrial development has been the heretofore cavalier disposal of halocarbons, whether used as degreasing solvents in metal processing, as non-flammable insulating oils in electrical transformers, or purposely used as defoliants, as with dioxin. Studies have shown that ionizing radiation from an electron beam can dehydrohalogenate these materials, leaving a less hazardous organic material and a soluble halide. Studies were conducted using a full-scale 1.5 MeV electron beam at a Miami–Dade waste treatment facility. Table V shows the effectiveness of this radiation process in remediating contaminated water. Lower voltage, self-shielded transportable electron beams were developed and placed in tractor-trailers for on-site demonstrations of soil and water remediation. While technically feasible, this use of radiation processing has not been a commercial success. Environmental engineers seem to be more familiar with alternative processes involving percolation, distillation, and other wet chemistry separation techniques (*8–10*).

Table V. Remediation of Contaminated Water

System: 375 to 600 L/min
CCl_2=CHCl in water at 100 mg/L
Electron beam dose = 5 kGy
Result: >99.9% removal of CCl_2=CHCl

Stack Gas Treatment

Countries dependent on coal as their primary fuel source, such as Poland, have investigated the use of electron beam treatment of stack gases to eliminate SO_2 and NO_x emissions from power plants. When irradiated in the presence of a small amount of ammonia, these undesirable flue gases form a white precipitate that can be removed from the gas stream and collected for other uses, such as fertilizer, as shown in Figure 5. A production facility has been constructed in Poland that is capable of treating 270,000 m^3/h of gas emissions, which is half the stack gas output of a medium-sized 60 MW coal-fueled power plant. A two-stage process vessel has been constructed that uses two 700 keV, 300 kW accelerators in tandem in each zone. Table VI presents data indicating the effectiveness of this irradiation process for removing undesirable air pollutants that contribute to the formation of acid rain. In excess of $20 million in funding for this project was facilitated by the International Atomic Energy Agency (*11*).

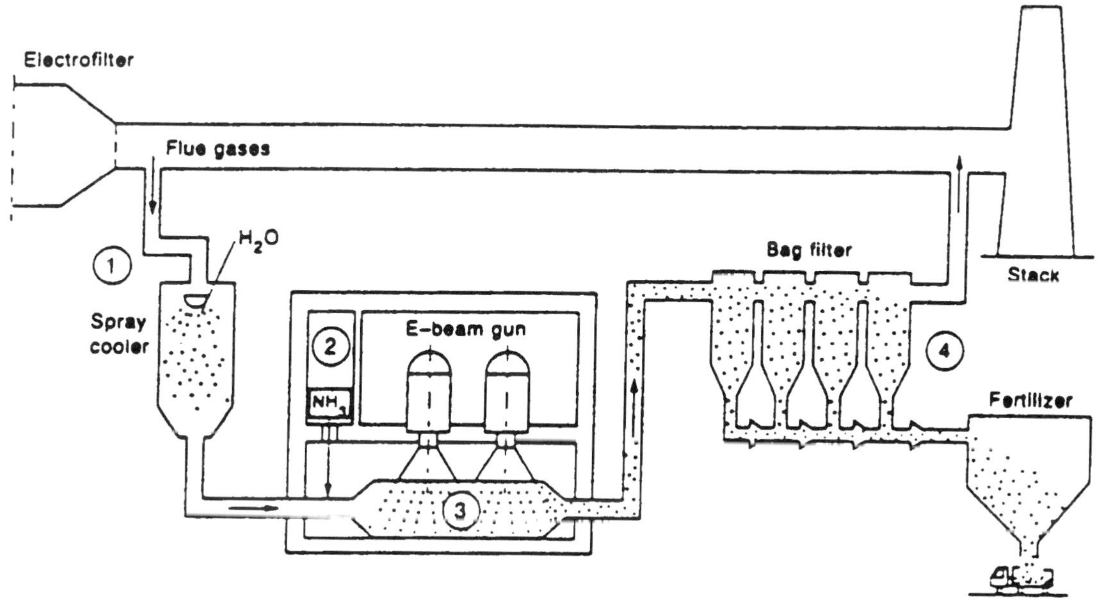

Figure 5. Stack gas irradiation facility.

Table VI. Irradiation Elimination of Stack Gas Contaminants

System: 135,000 m^3/h
SO_2 at 1.1 g/m^3 + NO_X at 0.7 g/m^3
9 kGy = 90% removal SO_2
= 70% removal NO_X

Summary

Radiation chemistry has formed the basis for a number of environmentally friendly or "green" industrial process technologies. Commercial success has been found in the elimination of volatile organic compounds for printing and coating processes. In these areas, high intensity ultraviolet light and low voltage electron beam curing and processing have proven to be point-of-source methods of pollution prevention processes, enabling companies to eliminate VOCs and meet EPA standards so as to comply with the Clean Air Act Amendments of 1990. This is complemented by significant improvements in overall process efficiency and energy savings. Manufacturers of diverse products, such as fiber optic cables, electronic circuit boards, adhesive coated materials, wood, and other coated products have adopted this mode of irradiation processing and its underlying chemistry.

While the technical feasibility has been demonstrated for using radiation chemistry in water, wastewater, and soil remediation, the implementation of these proven processes remains a challenge. However, scale up is underway involving the irradiation treatment of stack gases at a full-size coal-burning power plant in Europe.

Within the chemical industry, it has long been known that irradiation initiated synthesis can proceed at rapid rates with minimal need for catalysts. This can help to eliminate the problems of catalyst removal and disposal. Yet, this venue remains largely underexplored and has not been addressed by engineers familiar with chemical processes and reactor design.

Not only does the societal apprehension of the term "radiation" thwart some of these developments, but also a lack of inclusion of this form of energy transfer in most academic regimes and curricula hinders this even more. In the mid-1990s, a survey was conducted of some 51 universities that were considered to have an interest in radiation science and technology. This survey found that only six institutions had electron beam facilities that could even be considered for use in these environmental-type projects. Only a few more had industrially suited high-intensity ultraviolet light curing equipment. There existed but a few structured graduate programs involving photochemistry and irradiation science and technology. While outstanding work was being conducted through specific grants, it was found that there was neither a core curriculum nor a core text in this enabling area of technology. There was little to no provision for teaching radiation science and technology to undergraduate students (*12*).

In some instances, there has been an enlightened convergence of public support, mainly through state-based programs, for industrial implementation of environmentally friendly irradiation processing technologies, combined with public desires for pollution prevention and energy conservation. More such programs, with their supportive means to assist industry with capital conversion to irradiation processing, are needed.

Radiation chemistry will only prosper when there is greater public understanding of the benefits of this benign and environmentally friendly, "green" process. Radiation is, after all, as natural as sunlight itself.

References

1. Rechel, C. J., Ed. *UV/EB Curing Primer*; RadTech International North America: Bethesda, MD, 1991.
2. Bly, J. *Electron Beam Processing;* International Information Associates: Yardley, PA, 1988.
3. Berejka, A. J. Description and Use of High Current Industrial Electron Accelerators; *Enhancement of Wastewater and Sludge Treatment by Ionizing Radiation* (NSF Grant BES95-2059); University of Miami: Coral Gables, FL, 1997, pp 133–138.
4. Larsen, L. S.; Berejka, A. J. A Clean Sweep—UV Curing of Small Wood Products. *The RadTech Report* **1999**, (May/June), 20–23.
5. Berejka, A. J. Irradiation Processing in the '90's. *Radiation Physics and Chemistry* **1995**, *46* (4–6), 429–437.
6. Berejka, A. J. Reengineering PSA Manufacture for VOC Reduction and Energy Savings. *Adhesives Age* **1997,** (July), 30–36.
7. Berejka, A. J. Reengineering PSA Manufacture for VOC Reduction and Energy Savings. *Adhesives Age* **1997,** (Sept), 28–30.
8. Cooper, W. J.; Curry, R. D.; O'Shea, K. E., Eds. *Environmental Applications of Ionizing Radiation*; John Wiley & Sons: New York, 1998.
9. Nickelsen, M. G.; Kajdi, D. C.; Cooper, W. J.; Kurucz, C. N.; Waite, T. D.; Gensel, F.; Lorenzl, H.; Sparka, U. Field Application of a Mobile 20-kW Electron-Beam Treatment System on Contaminated Groundwater and Industrial Wastes. In *Environmental Applications of Ionizing Radiation*; Cooper, W. J., Curry, R. D., O'Shea, K. E., Eds.; John Wiley & Sons: New York, 1998, pp 451–466.

10. Schuetz, M. N.; Vroom, D. A. The Use of High-Power, Low-Energy Electron Beams for Environmental Remediation. In *Environmental Applications of Ionizing Radiation;* Cooper, W. J., Curry, R. D., O'Shea, K. E., Eds.; John Wiley & Sons: New York, 1998, pp 63–82.
11. Chmielewski, A. G.; Tyminski, B.; Iller, E.; Zimek, Z.; Licki, J. Electron-Beam Flue-Gas Treatment Process Upscaling. In *Environmental Applications of Ionizing Radiation;* Cooper, W. J., Curry, R. D., O'Shea, K. E., Eds.; John Wiley & Sons: New York, 1998, pp 197–216.
12. Berejka, A. J. Unpublished report to the Board of Directors of The RadTech Foundation, Ionicorp$^+$: Huntington, NY, January, 1996.

Chapter 13

Affordable Composites and Plastics from Renewable Resources: Part I: Synthesis of Monomers and Polymers

Richard P. Wool, Shrikant N. Khot, John J. LaScala, Shana P. Bunker, Jue Lu, Wim Thielemans, Erde Can, Shantaram S. Morye, and George I. Williams

Department of Chemical Engineering, ACRES Program, Center for Composite Materials, University of Delaware, Newark, DE 19716

Recent advances in genetic engineering, composite science, and natural fiber development offer significant opportunities for new, improved materials from renewable resources that can biodegrade or be recycled, thus enhancing global sustainability. A wide range of high-performance, low-cost materials can be made using plant oils, natural fibers, and lignin. By selecting the fatty acid distribution function of plant oils via computer simulation and the molecular connectivity, we can control chemical functionalization and molecular architecture to produce linear, branched, or cross-linked polymers. These materials can be used as pressure-sensitive adhesives, elastomers, rubbers, and composite resins. This chapter describes the chemical pathways that were used to modify plant oils and allow them to react with each other and various co-monomers to form materials with useful properties.

Introduction

Polymers and polymeric composites have extensive applications in the aerospace, automotive, marine, infrastructure, military, sports, and industrial fields. These lightweight materials exhibit excellent mechanical properties, high corrosion resistance, dimensional stability, and low assembly costs. Polymers traditionally are derived from petroleum, and as the number of applications increases, alternative sources of these materials become critical. In recent years, the Affordable Composites from Renewable Sources (ACRES) program at the University of Delaware has developed a broad range of chemical routes to use natural triglyceride oils as a feedstock for polymers and composites (*1*, *2*). The environmental and economic advantages of these materials make them attractive alternatives to petroleum-based materials.

Natural oils from both plant and animal sources are abundant in most parts of the world, making them an ideal alternative to chemical feedstocks. These oils are predominantly made up of triglyceride molecules, which have three fatty acids joined at a glycerol juncture, Structure 1. The fatty acids in the most common oils are 14 to 22 carbons long with 0 to 3 double bonds per fatty acid. Exotic oils are composed of fatty acids with other types of functionalities, such as epoxies, hydroxyls, cyclic groups, and furanoid groups (*4*). On a molecular level, these oils are composed of many different types of triglycerides with numerous levels of unsaturation. Newly developed genetic engineering techniques can control the variation in unsaturation in plants such as soybean, flax, and corn, but some oils are better suited to polymer resin development.

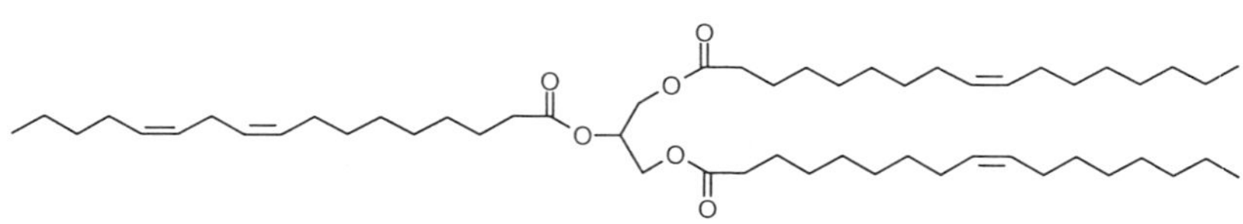

Structure 1

Besides applications in the foods industry, triglyceride oils have been used extensively to produce coatings, plasticizers, lubricants, agrochemicals, and inks (*5*, *11*). In polymers, these oils can be used as toughening agents and to produce interpenetrating networks that can increase the toughness and fracture resistance in conventional thermoset polymers (*12–25*). In these works, the functional triglyceride was a minor component in the polymer matrix acting solely as a modifier to improve the physical properties of the main matrix. Consequently, these low-M_W, lightly cross-linked materials by themselves did not have the rigidity and strength required for structural applications.

Synthetic Pathways for Triglyceride-Based Monomers

Triglycerides contain active sites amenable to chemical reaction (the double bond, allylic carbons, the ester group, and carbons alpha to the ester group) that can be used to introduce polymerizable groups by using the same techniques applied in the synthesis of petrochemical-based polymers. The key step is reaching a higher M_W and cross-link density and incorporating chemical functionalities known to impart stiffness in a polymer network (e.g., aromatic or cyclic structures). Figure 1 illustrates several possible synthetic pathways (*1*).

In **5–8** and **11** the double bonds functionalize the triglyceride with polymerizable chemical groups. From the natural triglyceride, it is possible to attach maleates (**5**) (*6–11*) or to convert the unsaturation to epoxy (**7**) (*26–28*) or hydroxyl functionalities (**8**) (*29, 30*). Such transformations allow the triglyceride to react via ring-opening or polycondensation polymerization. Vinyl functionalities can be attached by reacting the epoxy functional triglyceride with acrylic acid to incorporate acrylates (**6**) and by reactiing the hydroxyl functional triglycerides with maleic anhydride to incorporate maleate half-esters and esters (**11**). These monomers can then be blended with a reactive diluent, similar to most conventional vinyl ester resins, and cured by free-radical polymerization.

The second method for synthesizing monomers is to convert the triglyceride to monoglycerides through a glycerolysis (**3A**) reaction or an amidation reaction (**2, 3B**) (*31–36*). Monoglycerides, commonly referred to as alkyd resins, are used in surface coatings because of their low cost and versatility (*32*). In those applications the double bonds of the monoglyceride react to form the coating, but monoglycerides also can react through the alcohol groups via polycondensation reactions with a co-monomer, such as a diacid, epoxy, or anhydride. Alternatively, maleate half-esters can be attached to the monoglycerides (**9**), allowing them to undergo free-radical polymerization.

The third method is to functionalize the unsaturation sites as well as reduce the triglyceride into monoglycerides. This can be accomplished by glycerolysis of an unsaturated triglyceride, followed by hydroxylation or glycerolysis of a hydroxy functional triglyceride. The product can then react with maleic anhydride to form a monomer capable of free-radical polymerization (*1*).

The structure of triglycerides is complex, but some aspects can be characterized with proton nuclear magnetic spectroscopy (^{1}H NMR) and Fourier transform infrared (FTIR) spectroscopy. A typical ^{1}H NMR spectrum of soybean oil is shown in Figure 2, with peak assignments. The two sets of peaks at 4.0 to 4.4 ppm are produced by the four glycerol methylene protons per triglyceride (*4*). The triplet set of peaks at 2.3 ppm is produced by the six protons in the alpha position relative to the carbonyl groups. The peak at 0.9 ppm is produced by the 9 methyl protons per triglyceride at the end of each fatty acid chain. These three groups of peaks provide a standard by which other peaks can be used to quantitatively characterize functional groups in the triglyceride.

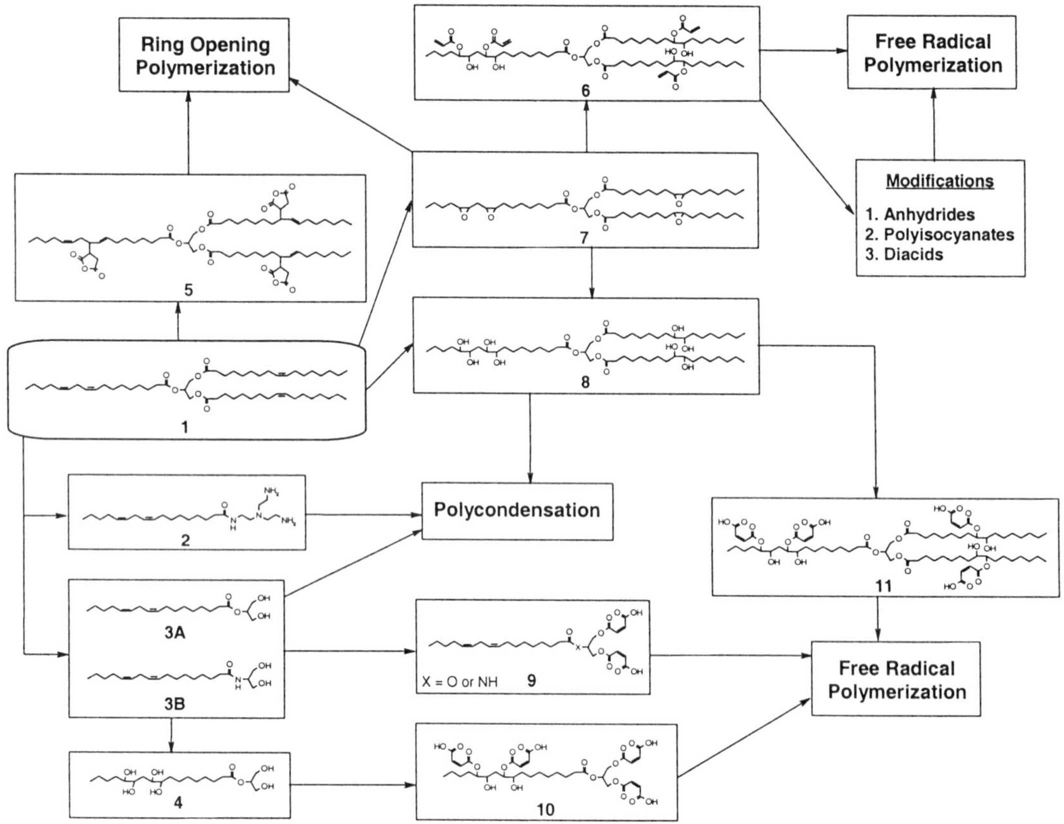

Figure 1. Chemical pathways leading to polymers from triglyceride molecules (1).

In this work we focus on three triglyceride monomers; acrylated epoxidized soybean oil (AESO, structure **12A**), maleinized soybean oil monoglyceride (SOMG/MA, structure **12B**), and maleinized hydroxylated soybean oil (HSO/MA, structure **12C**). These are promising candidates for use in the composites and engineering plastics fields. When used as a major component of molding resin, these monomers have shown properties comparable to conventional polymers and composites, as will be presented here. They can also be used as a matrix in synthetic and natural fiber reinforced composites.

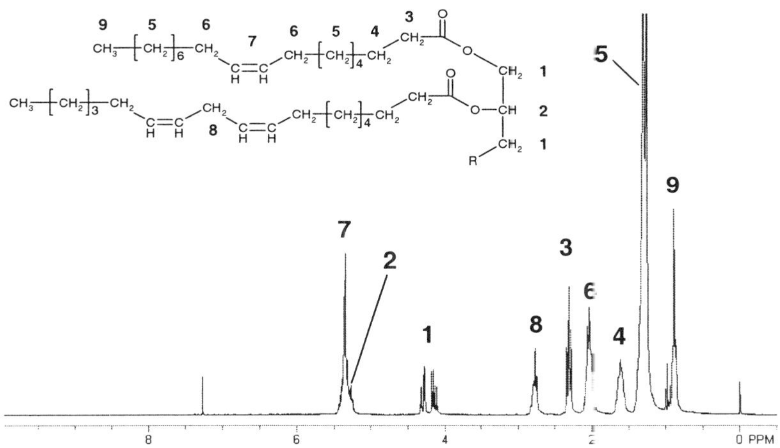

Figure 2. ^{1}H-NMR spectrum of soybean oil. R represents a third fatty acid.

Acrylated Epoxidized Soybean Oil

Acrylated epoxidized oils are synthesized from the reaction of acrylic acid with epoxidized triglycerides, which can be found in natural oils such as vernonia plant oil or can be synthesized from more common unsaturated oils, such as soybean or linseed oils (*37*). Vernonia oil has a functionality of 2.8 epoxy rings per triglyceride (*13*). Epoxidized soybean oil (ESO) is commercially available and is generally sold with a functionality of 4.1 to 4.6 epoxy rings per triglyceride, which can be identified via ^{1}H NMR (*20*, *38*). Epoxidized linseed oil is also commercially available when higher epoxy content is required. Predominantly, these oils are used as alternative plasticizers in polyvinyl chloride in place of phthalates (*39–41*), but their use as a toughening agent also has been explored (*20*, *23–25*, *42*). With the addition of acrylates, the triglyceride can be reacted via addition polymerization. AESO has been used extensively in surface coatings and is commercially made in forms such as

Ebecryl 860 (*7, 43, 44*). Urethane and amine derivatives of AESO have also been developed for coating and ink applications (*8, 9, 45*).

Acrylic acid and ESO react via a standard substitution reaction that has 1st-order dependence with respect to epoxy concentration and 2nd-order dependence with respect to acrylic acid concentration (*46*). However, epoxidized oleic methyl ester has a 2nd-order dependence on both epoxy and acrylic acid concentrations (*47*). Although the reaction of ESO with acrylic acid is partially catalyzed by the acrylic acid, the use of additional catalysts is common. Tertiary amines, such as N,N-dimethyl aniline, triethylamine, and 1,4-diazobicyclo[2.2.2]octane, are commonly used (*38, 48*). Additionally, more selective organometallic catalysts have been developed that reduce the amount of epoxy homopolymerization (*49, 50*).

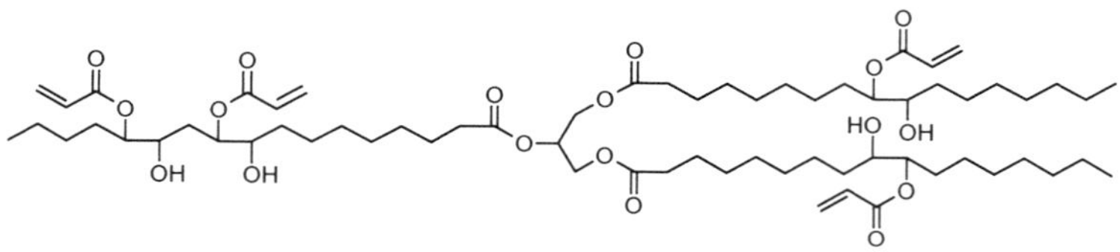

Acrylated Epoxidized Soybean Oil (AESO)

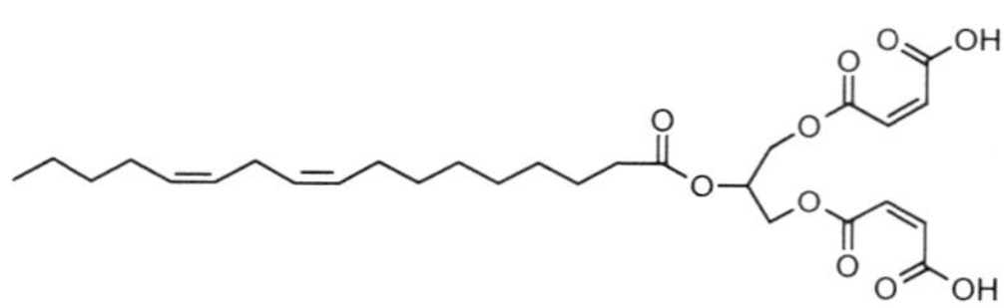

Maleinated Soybean Oil Monoglyceride (SOMG/MA)

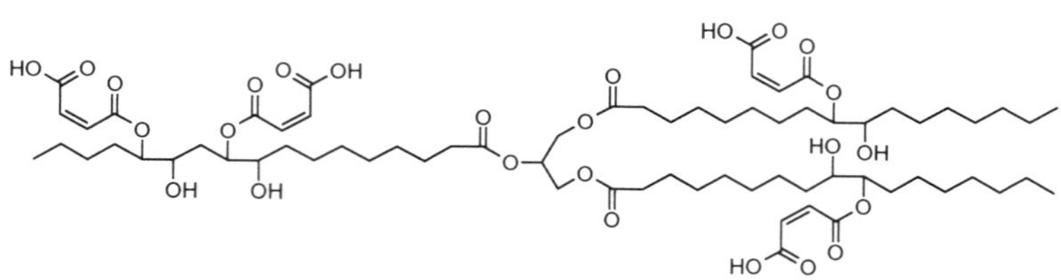

Maleinated Hydroxylated Soybean Oil (HSO/MA)

Structures 12A (top), 12B (middle), 12C (bottom)

AESO can be blended with a reactive diluent such as styrene to improve its processability and to control the polymer properties (e.g., modulus, glass transition temperature) to reach a range acceptable for structural applications. Polymer properties can also be controlled by changing the M_W of the monomer or the functionality of the acrylated triglyceride. Consequently, a range of properties, and therefore applications, can be found. After the acrylation reaction, the triglyceride contains residual amounts of unreacted epoxy rings as well as newly formed hydroxyl groups, both of which can be used to further modify the triglyceride by reaction with a number of chemical species, such as diacids, diamines, anhydrides, and isocyanates. The approach presented here is to oligomerize the triglycerides with reagents that have chemical structures conducive to stiffening the polymer, such as cyclic or aromatic groups. Reacting AESO with cyclohexane dicarboxylic anhydride (CDCA) forms oligomers (Scheme A), increasing the entanglement density as well as introducing stiff cyclic rings to the structure. Reacting AESO with maleic acid (Scheme B) also forms oligomers as well as introduces more double bonds. Although it is desirable to maximize the conversion of hydroxyls or epoxies, viscosity increases dramatically at high levels of conversion. Eventually, this can lead to gelation, so the reaction must be carefully monitored. After oligomerization, the modified AESO resin can be blended with styrene and cured in the same manner as unmodified AESO resin.

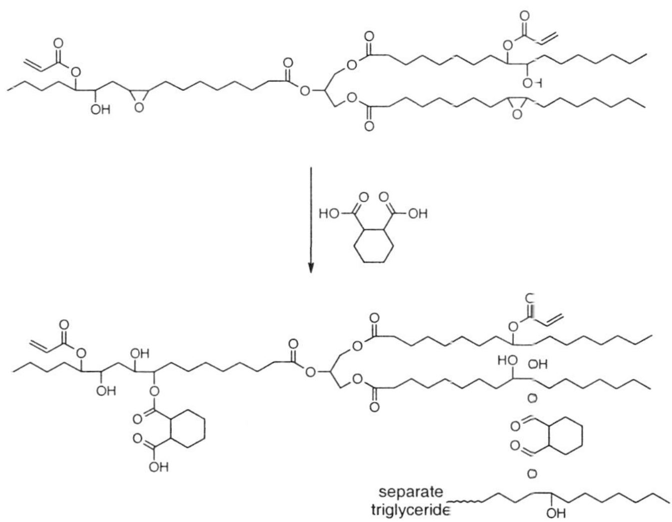

Scheme A. Modification of AESO by reaction with CDCA.

Maleinized Soyoil Monoglyceride

Maleinized soyoil monoglyceride (**12B**) is synthesized from the triglyceride oil in two steps (*33*). The first is a standard glycerolysis reaction to convert the triglycerides into monoglycerides by reaction with glycerol; see (*31*). The product is generally a mixture of mono- and diglycerides, Scheme C. Using excess glycerol can aid in conversion, and the reaction can be run in solvent or in the presence of an emulsifier catalyst (*34*). Once the reaction is completed, it is possible to separate a portion of the unreacted glycerol by cooling the product rapidly (*33*). The presence of glycerol is not detrimental, because it can be reacted with maleic anhydride in the same manner as the monoglycerides and incorporated into the end polymer network.

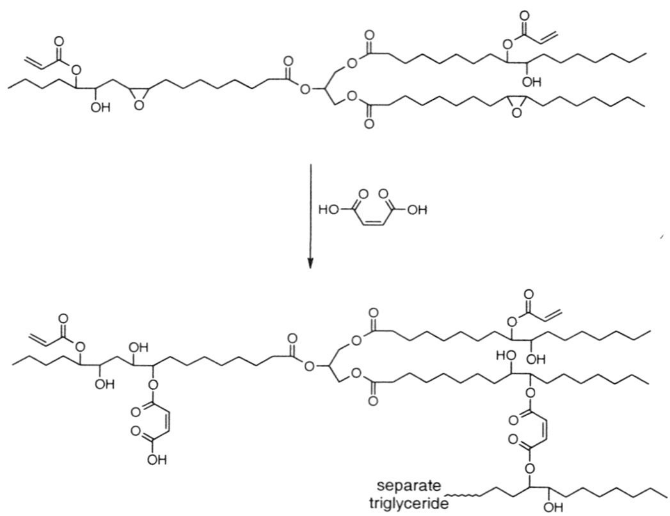

Scheme B. Modification of AESO by reaction with maleic acid.

Maleinization of the SOMG mixture at temperatures (*t*'s) <100 °C produces monoglyceride, diglyceride, and glycerol maleate half-esters. This reaction does not attempt to produce a polyester; half-ester formation is expected to proceed at low *t* in the presence of either acid or base catalysts without any byproducts. A good indication of reaction success is the signal intensity ratio of maleate vinyl protons to fatty acid vinyl protons ($N_M \backslash N_{FA}$) in the ^{1}H NMR spectrum. The use of 2-methylimidazole and triphenyl antimony as catalysts has been successful at reaction *t*'s of 80 to 100 °C with a 3:2 weight (wt) ratio of glycerides to maleic anhydride ($N_M \backslash N_{FA}$=0.85) (*33, 51*). Once these maleates are added, the monoglycerides can react via addition polymerization. Maleates are relatively unreactive with each other, so adding styrene increases the polymerization conversion and imparts rigidity to the matrix.

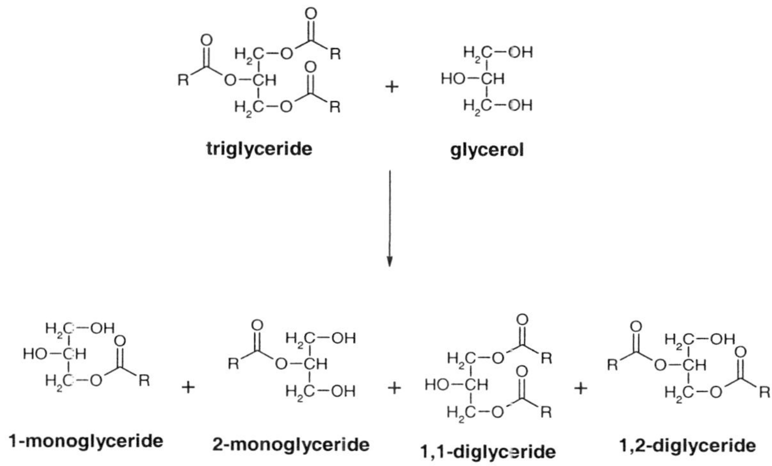

Scheme C. Glycerolysis of **1** *to form a mixture of mono- and diglycerides.*

To increase the glass transition temperature (T_g) and modulus of the SOMG/MA polymer, more rigid diols, such as neopentyl glycol (NPG) and bisphenol A (BPA), can be added during the maleinization reaction to increase the rigidity of the end polymer network. Although this will reduce the renewable resource content of the final resin, the T_g of the end polymer should be higher. The synthesis of maleate half-esters of organic polyols, including NPG and BPA and the cross-linking of the resulting maleate half-esters with a vinyl monomer such as styrene, has been reported (*52, 53*). The literature provides many examples of unsaturated polyesters prepared from NPG and maleic anhydride with other polyols and diacids (*54-57*). However, the coploymers of NPG and BPA bis maleate half-esters with SOMG maleate halfesters is new.

Here we present the properties of the SOMG/MA polymer as well as the effect of adding NPG and BPA on the mechanical properties of the final polymers. For this purpose, mixtures of SOMG/NPG and SOMG/BPA prepared at the same wt ratio were maleinized, and the mechanical properties of the copolymers of the resulting maleates with styrene were analyzed and compared to those of SOMG maleates.

Maleinized Hydroxylated Oil

The maleinized hydroxylated oil (HO/MA) synthesis is similar to that of AESO and SOMG/MA. The double bonds of an unsaturated oil are used to attach polymerizable groups by converting the double bonds of the triglyceride to hydroxyl groups, which can then be used to attach maleates. As shown in Figure 1, there are two routes to synthesize the hydroxylated triglyceride. The first is through an epoxidized intermediate; by reacting the epoxidized triglyceride with an acid, the epoxies can be easily converted to hydroxyl groups (*29, 58*). Alternatively, the hydroxylated oil can be synthesized directly from the unsaturated oil (*30*) and then be reacted with maleic anhydride to functionalize the triglyceride with maleate half-esters. A molar ratio of 4:1 (anhydride to triglyceride) was used in all cases, and the reaction was catalyzed with N,N-dimethylbenzylamine. Once the maleinization reaction is finished, the monomer resin can be blended with styrene, similar to the other resins presented here.

Experimental

Acrylated Epoxidized Soybean Oil Polymers

AESO was examined for its ability to produce high-T_g and high-modulus polymers. Ebecryl 860, a commercial form of AESO, was blended with various amounts of styrene to determine the effects on mechanical and dynamic mechanical properties. The average functionality of the AESO was about 3 acrylates per triglyceride as determined by ^{1}H NMR (*38*); an example spectrum is shown in Figure 3. Similar to soybean oil, the triplet peak at 2.3 ppm can be used as a basis for the protons present alpha to the carbonyls in the triglyceride. The three peaks in the range of 5.8 to 6.5 ppm represent the three protons of the acrylate group.

Styrene monomer was blended with AESO and a free radical initiator, 2,5-dimethyl-2,5-di(2-ethylhexanoyl peroxy) hexane. Adding styrene to any type of unsaturated polyester is common practice in the composite liquid molding resin field. Its low cost and low viscosity improve the price and processability of the resin. For triglyceride-based polymers, styrene also imparts a rigidity that the triglyceride does not naturally possess. The amount of initiator used was 1.5 wt % of the total resin wt (AESO plus styrene). For tensile testing of the polymers, samples were prepared per ASTM D 638. The resin was cured at 60 °C for 12 h, followed by 125 °C for 1.5 h. Samples for dynamic mechanical analysis (DMA) testing were prepared by pouring resin into a rubber gasket between two metal plates covered with aluminum foil. Samples were cured at 65 °C for 1.5 h and post-cured at 125 °C for 1.5 h.

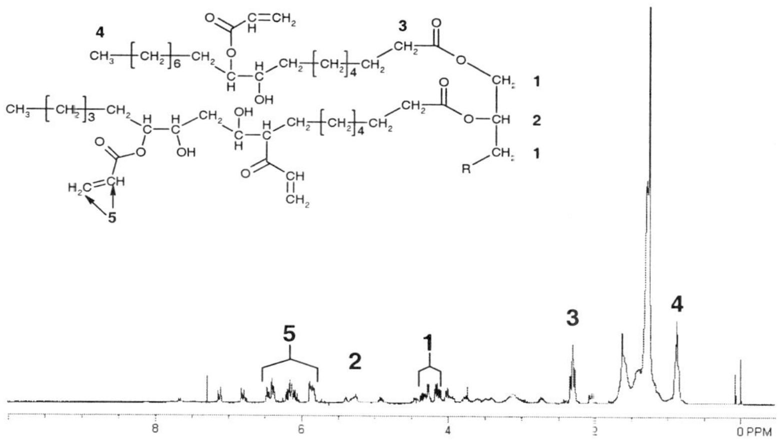

Figure 3. ^{1}H-NMR spectrum for AESO (Ebecryl 860).

Tensile properties of the polymer were determined with an Instron 4502 with Series II automated materials testing system v.4.05e. Samples were tested at a cross-head speed of 0.02 in/min. Strain measurements were taken using biaxial strain gages and Labview Data Analysis software. DMA was conducted in a three-point bending geometry on a Rheometrics Solids Analyzer II. Temperature was ramped from –125 to 120 °C at a rate of 5 °C/min, with a frequency of 1 Hz and strain of 0.01%.

Synthesis of Modified Acrylated Epoxidized Soybean Oil Polymers

To improve the properties of AESO-based resins, modified forms of AESO were synthesized. These modifications involved partially reacting ESO with acrylic acid and reacting the remaining epoxies with anhydrides or diacids. A more detailed explanation of the synthesis of partially acrylated ESO can be found in other sources (*38*). In summary, a mixture of ESO was mixed with a stoichiometric amount of acrylic acid (about 1500 g ESO to 460 g acrylic acid). Hydroquinone was added as a free radical inhibitor in the amount of 0.07% of the total reactants' wt, as well as 1,4-diazobicyclo[2.2.2]octane to act as a catalyst in the amount of 0.1% by wt. This was reacted at 95 °C for about 11 h and then allowed to cool to room *t*. The resulting product had about 1.7 acrylates/triglyceride and 0.4 residual epoxies/triglyceride, according to ^{1}H NMR. The remaining 2.3 epoxies were lost to epoxy homopolymerization (*38*).

The first modification was the reaction of AESO with CDCA (**A**). In a typical reaction, the synthesized AESO was reacted with 7.4% of its wt in CDCA and 0.1% of its wt in 2-methyl imidazole, which catalyzes the reaction (*38*).

After reacting at 110 °C for about 3 h, the majority of anhydride and epoxy groups were consumed, as indicated by FTIR spectroscopy. The second modification was the reaction of AESO with maleic acid, **B**. This was accomplished by reacting the synthesized AESO with 11% of its wt in maleic acid (*38*). The reaction was held at approximately 80 °C for 4 h, during which consumption of the epoxies was again confirmed by FTIR spectroscopy.

The modified resins were then blended with styrene and initiator in the amounts of 66 wt % modified AESO, 33 wt % styrene, and 1 wt % 2,5-dimethyl-2,5-di(2-ethylhexanoyl peroxy) hexane initiator. After curing at 65 °C for 1.5 h and post-curing at 125 °C for 1.5 h, the polymers' dynamic mechanical properties were analyzed and compared to unmodified AESO resin.

Maleinized Soyoil Monoglyceride Resin Synthesis

SOMG/MA was synthesized by breaking triglycerides into monoglycerides and then functionalizing the alcohol groups with maleic anhydride. The glycerolysis reaction was done by heating the triglycerides in the presence of glycerol and a catalyst. In a typical reaction, glycerol was heated at 220 to 230 °C for 2 h under an N_2 atmosphere to distill off any water (*33*). The reaction ratio was 4 g soybean oil to 1 g glycerol, a molar ratio of 4.75:1 (glycerol to triglyceride). The soybean oil was added in five portions to the glycerol, each 1 h apart. With the first portion commercial soap was added in the amount of 1% of the total oil amount to act as an emulsifier and catalyst. The solution was heated at 230 °C under N_2 while being stirred. After 5.5 h, the reaction was immediately cooled to room *t* with an ice bath, causing glycerol to separate from the mixture. Removal of this layer recovered about 90% of the reaction solution, consisting of glycerides and glycerol.

Maleinization of the mixture was accomplished by heating 60 g of glyceride/glycerol mixture to about 80 °C while being stirred. Forty g maleic anhydride was then added, and as it melted, 0.6 g triphenyl antimony was added as a catalyst along with 0.01 g hydroquinone. The reaction was complete after 5.5 h, according to FTIR and ^{1}H NMR, resulting in a mixture of maleinized glycerides and glycerol (SOMG/MA) (*33*).

Maleinized Soyoil Monoglyceride/Neopentyl Glycol Resin Synthesis

Modifying the procedure given in (*53*), SOMG/NPG/MA resin was synthesized by placing 45 g of SOMG into a 250-mL round-bottom flask equipped with a *t* controller and magnetic stirrer and heating to 125 °C (*51*). Fifteen g NPG (0.144 mol) was then added, and as the NPG melted, 58.3 g maleic anhydride was added. As the three compounds formed a homogenous solution, 0.06 g triphenyl antimony catalyst and 0.015 g hydroquinone were added. The solution was stirred for 6.5 h at 120 °C. ^{1}H NMR analysis of the product showed the formation of both the SOMG and NPG maleate and later fumarate vinyl groups. The product was a light yellow viscous liquid at room *t*.

Maleinized Soyoil Monoglyceride/Bisphenol A Resin Synthesis

The preparation of maleates of BPA and ethylene and propylene oxide adducts of BPA has been reported (*52*). For this work, SOMG and BPA were maleinized as a mixture (*51*); 45 g SOMG was placed into a 250-mL round-bottom flask equipped with a *t* controller and a magnetic stirrer and heated to 125 °C. Fifteen g BPA (0.0657 mol) was added, and as that dissolved, 42.88 g maleic anhydride (0.4375 mol) was added. As the three compounds formed a homogenous solution, 0.6 g triphenyl antimony and 0.01 g hydroquionone were added. The solution was then stirred for 9 h at 125 °C until maleic anhydride consumption was over. ^{1}H NMR analysis of the product showed the formation of both SOMG and BPA maleate and later fumarate vinyl groups. The reaction product was an orange-colored viscous liquid (98 g) at room *t*.

Copolymerization of the Maleates with Styrene

Copolymerizations of SOMG/MA, SOMG/NPG/MA, and SOMG/BPA/MA with styrene were all run under the same conditions to compare the mechanical properties of the resulting polymers. For this purpose a certain wt ratio of the maleate mixture was mixed with 35% of its own wt of styrene in a closed vial. All of the maleate products were soluble in styrene. Tert-butyl peroxy benzoate radical initiator, 2% by wt of the total mixture, was added. Nitrogen gas sparging and vacuum degassing were carried out for 5 min. The solution was then transferred to a rectangular rubber gasket mold sandwiched between two steel plates. The resin-filled mold was heated to 120 °C at a rate of 5 °C/min, cured at this *t* for 3.5 h, and then post-cured at 150 °C for 1 h. Samples were clear, homogenous, and free of voids or gas bubbles. The polymer samples were polished and prepared for DMA, which was conducted in a three-point bending geometry on a Rheometrics Solids Analyzer II. Temperature was ramped from 30 to 200 °C at 5 °C/min, with a frequency of 1 Hz and strain of 0.01%.

Maleinated Hydroxylated Oil Polymer Synthesis

HO/MA (**12C**) uses the unsaturation of the triglyceride to incorporate polymerizable groups. This monomer was used in a series of experiments to understand how triglyceride structure can affect the synthesis and dynamic mechanical properties of the end polymer (*59*). Olive oil, cottonseed oil, soybean oil, safflower oil, linseed oil, triolein, and a genetically engineered high oleic soybean oil were converted into HO/MA resins. The fatty acid chain lengths for all of these oils are between 17.5 and 18 carbons, making the unsaturation level essentially the only difference among oils.

Hydroxylation was done by stirring the oil (~100 g) vigorously in the presence of formic acid (150 mL) and 30% (aq) hydrogen peroxide (55 mL) at 25 °C (*30, 59*). The reaction time was 18 h to reach a maximum conversion of double bonds. Formic acid, peroxide, and water were then removed from the hydroxylated oil by dissolving the reaction mixture in diethyl ether and washing multiple times with water and then saturated (aq) sodium bicarbonate until reaching a neutral pH. The solution was then washed with saturated (aq) sodium chloride and dried over sodium sulfate. Finally, the ether was evaporated off under vacuum. The extent of hydroxylation can be characterized by ^{1}H NMR. An example ^{1}H NMR spectrum is presented in Figure 4 with corresponding peak assignments (*59*). The extent of hydroxylation has a linear dependence on the level of unsaturation. Generally, for every double bond present on the triglyceride, an average of 1.6 hydroxyls can be added (*59*).

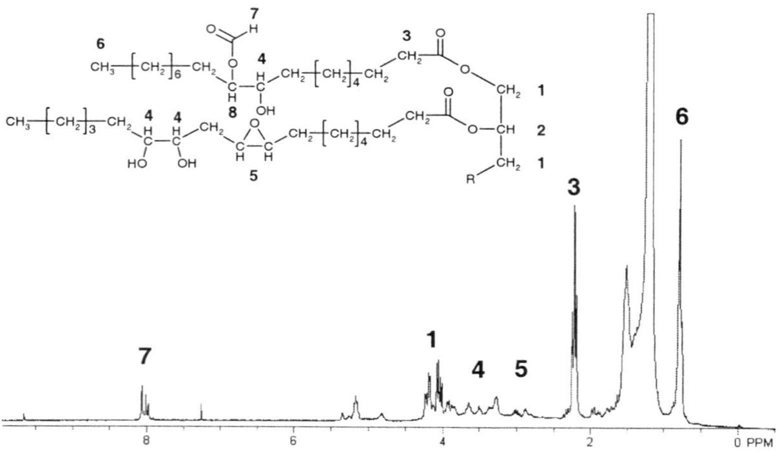

Figure 4. ^{1}H-NMR of hydroxylated soybean oil. Treating the oil with formic acid and hydrogen peroxide converts the double bonds to hydroxy groups.

The purified hydroxylated oil was reacted with maleic anhydride in a molar ratio of 1:4 (triglyceride:anhydride). The oil was heated to about 80 °C, and finely ground maleic anhydride was then added. After the anhydride dissolved N,N-dimethylbenzylamine was added to catalyze the reaction. The reaction was continued for 3 h, and the extent of maleinization was determined by ^{1}H NMR, as illustrated in Figure 5 (*59*). Under these reaction conditions, the extent of functionalization plateaus in the range of 2.1 to 2.8 maleates per triglyceride for all oils (*78*). About 20 to 25% of the maleates attached to the triglycerides isomerize to form fumarate groups (trans confirmation). Unreacted maleic anhydride remained in the resin and was polymerized during the cure reaction. The HOMA resins were then dissolved in styrene in a molar ratio of 7:1 (styrene:HOMA). The resins were cured at 65 °C for 1.5 h using 2,5-dimethyl-2,5-di-(2-ethylhexanoylperoxy)hexane and post-cured at 120 °C for 1 h. DMA was conducted in a three-point bending geometry on a Rheometrics Solids Analyzer II. Temperature was ramped from 30 to 175 °C at a rate of 5 °C/min, with a frequency of 1 Hz and strain of 0.01%.

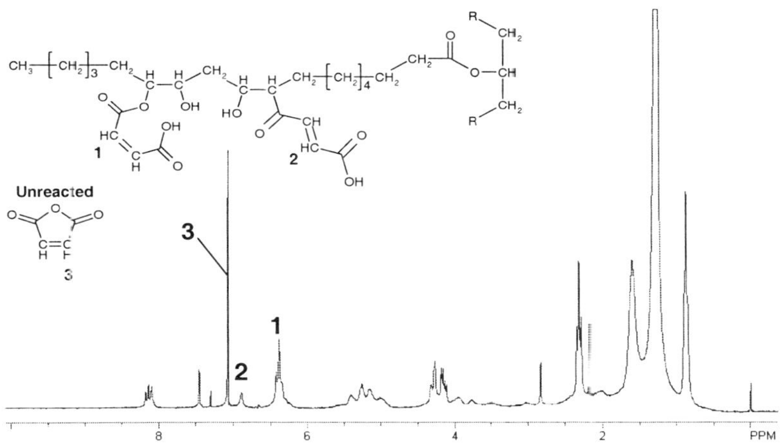

Figure 5. ^{1}H-NMR of maleinized hydroxylated soybean oil. Peaks 1 and 2 represent the maleate half esters and fumarate half-esters, respectively. Peak 3 represents unreacted maleic anhydride.

Results and Discussion

Viscoelastic and Mechanical Properties of AESO/Styrene Polymers

The storage moduli, E', of the AESO/styrene neat polymers at various *t*'s and compositions are shown in Figure 6. At room *t*, E' is proportional to the amount of styrene present, which is expected from the tensile properties presented earlier. Additionally, at room *t* all of the polymers are in the transition phase from the glassy region to the rubbery plateau. Even at *t*'s as low as –130 °C, these polymers have not reached a characteristic glassy plateau. At extremely low *t*'s, all compositions exhibit essentially equal moduli of about 4 GPa. At higher *t*'s, the compositions show moduli inversely proportional to the amount of styrene present. According to rubber elasticity theory (*60*), the lower styrene content polymers have a higher cross-link density, as seen in Figure 6.

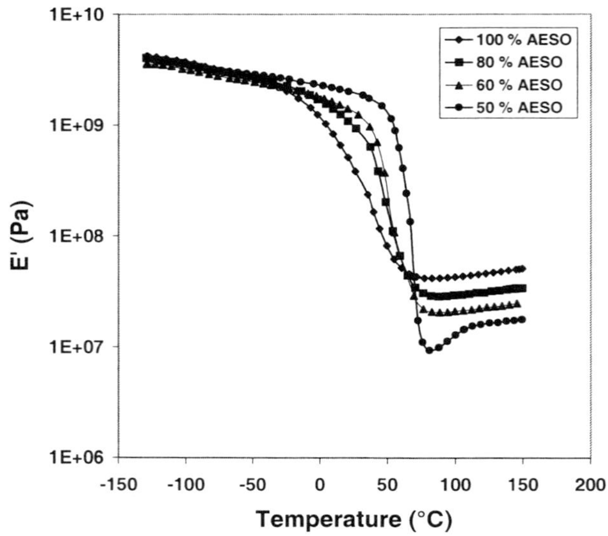

Figure 6. Storage modulus (E') of AESO-styrene copolymer as a function of t.

The T_g is often designated as either the *t* at which the dynamic loss modulus E" value is at a peak or the *t* at which the loss tangent tan δ exhibits a peak (*61*). As shown in Figure 7, all of the AESO-styrene copolymers exhibit two peaks in E". A minor relaxation occurs in the range of -85 to -95 °C, showing little dependence on composition. The much larger relaxation, corresponding to the T_g, occurs in the range of -10 to 60 °C and also becomes sharper in nature with the addition of styrene. These peaks are shown in the tan δ graph in Figure 8. The *t* at which these peaks occur exhibits a linear dependency on composition, increasing with the amount of styrene present in the system, as illustrated by Figure 9.

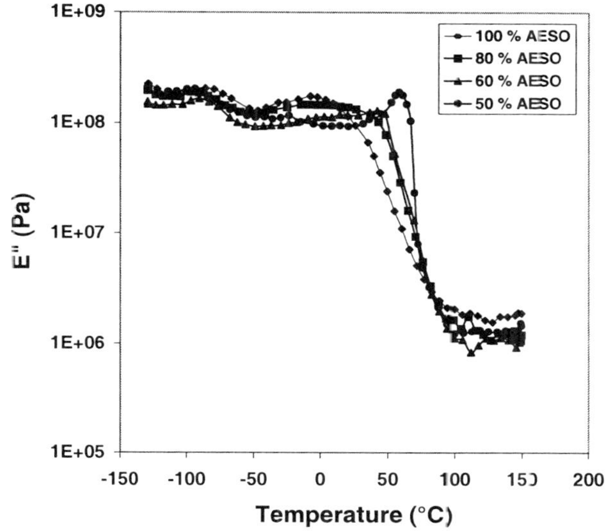

Figure 7. Loss modulus (E") of AESO-styrene copolymer as a function of t.

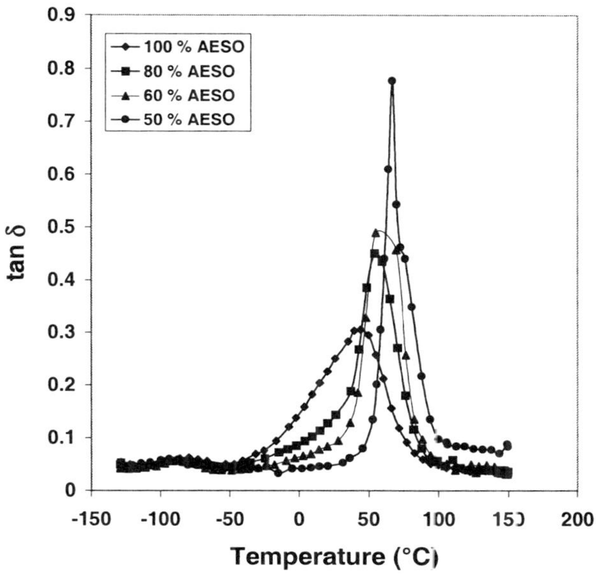

Figure 8. Dampening peak (tan δ) of AESO-styrene copolymer as a function of t.

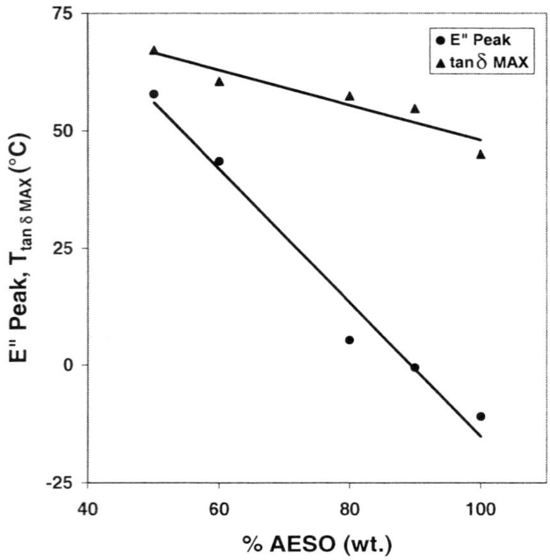

Figure 9. E" and tan δ peak temperatures of various compositions of AESO-styrene polymer.

The above dynamic mechanical behavior is a combination of three factors, cross-link density, copolymer effects, and plasticization. As the amount of AESO increases, so does the number of multifunctional monomers. Therefore, the overall cross-link density will be greater with increasing amounts of AESO, as supported by the high-*t* moduli shown in Figure 6. Increasing the cross-link density slows the transition in E' from glassy to rubbery behavior, and the tan δ peak broadens and decreases in height (*61*). The copolymer effect occurs frequently when there are differences in the reactivity or structure of the different monomers. If one monomer is more reactive, it is depleted faster, causing polymer formed later in the reaction to be composed mostly of the slower-reacting monomer. This causes heterogeneity in the composition of the total polymer. If these monomers differ in their physical properties, such as very different T_g's, a general broadening of the glass-rubber transition is frequently observed because of this gradient (*61*).

The other factor in the dynamic mechanical behavior, plasticization, is due to the molecular nature of the triglyceride. The starting soybean oil contains fatty acids that are completely saturated and cannot be functionalized with acrylates. These fatty acids act as a plasticizer, introducing free volume and enabling the network to deform more easily. The addition of even small amounts of plasticizer to polymers has been known to drastically broaden the transition from glassy to rubbery behavior and reduce the overall modulus (*61*). This effect may be inherent to all natural triglyceride-based polymers that use the double bonds to add functional groups. However, with advances in genetic engineering it may be possible to reduce this trend by reducing the amount of saturated fatty acids present, thus sharpening the glass-rubber transition. This issue is addressed later in the properties of HO/MA polymers produced from genetically engineered high oleic content oil and synthetic triolein oil. The existence of some saturated fatty acids, though, can contribute to improved toughness and ballistic impact resistance (*62*).

Tensile Properties of AESO/Styrene Polymers

The tensile moduli of three AESO-styrene copolymers at room *t* are shown in Figure 10. The pure AESO polymer has a modulus of about 440 MPa. At a styrene content of 40 wt % the modulus increases 4-fold to 1.6 GPa. In this region the dependency on composition is fairly linear. The ultimate tensile strengths of these materials, as shown in Figure 11, also show linear behavior. The pure AESO samples exhibited strengths of about 6 MPa, while the polymers with 40 wt % styrene show much higher strengths of about 21 MPa. The addition of styrene drastically improves the properties of the end resin.

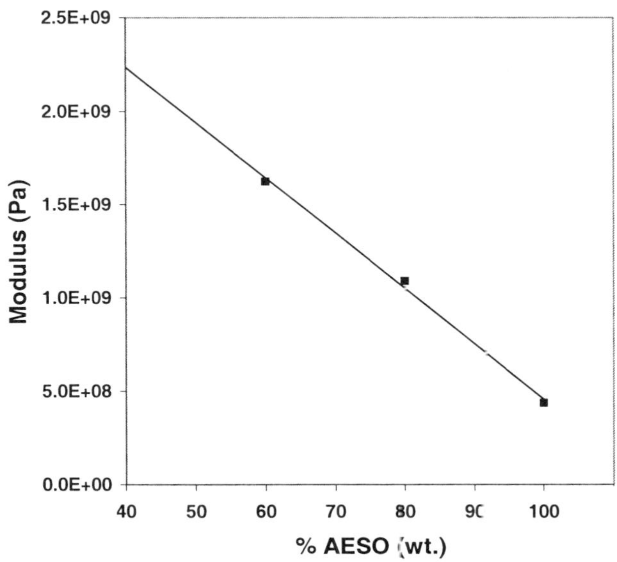

Figure 10. Tensile modulus of AESO-styrene copolymers.

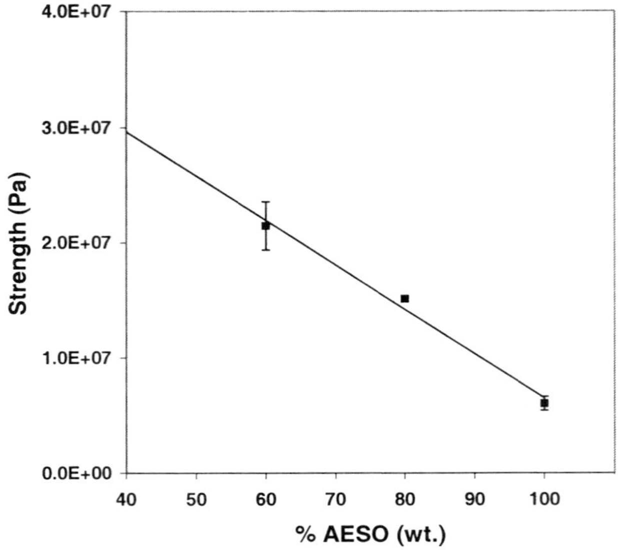

Figure 11. Ultimate tensile strength of AESO-styrene copolymers.

Dynamic Mechanical Behavior of Modified AESO Resins

The dynamic mechanical properties of the AESO polymers modified by CDCA and maleic acid were better than the unmodified polymers. As shown in Figure 12, E' at room *t* increases with both of these modifications. The E' of the unmodified AESO resin at room *t* is 1.3 GPa, while the CDCA modification increases the modulus to 1.6 GPa. The maleic acid modification provides the most improvement, raising E' to 1.9 GPa. The T_g, as indicated by the peak in tan δ, does not show any large increase from the anhydride modification, as shown in Figure 13. However, the maleic acid modification shifts the tan δ peak by almost 40 °C, showing a peak at 105 °C. The increased broadness of the peak can be attributed to increased cross-link density.

SOMG/MA Polymer Properties

As seen in Figure 14, the tan δ peak for the SOMG/MA polymer occurs at around 133 °C, and the polymer has an E' of about 0.92 GPa at room *t*. The

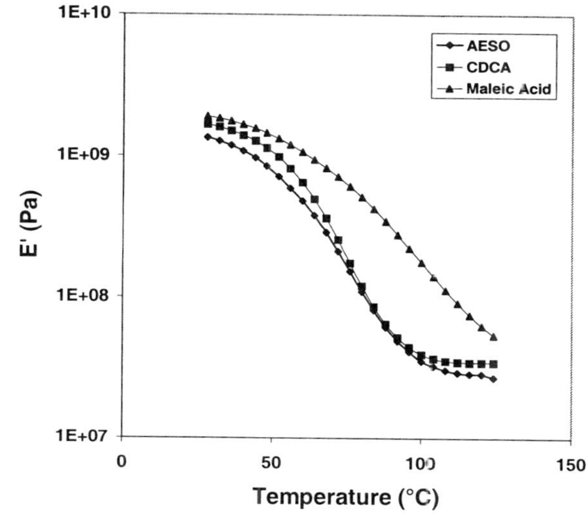

Figure 12. E' of modified AESO resins as a function of t.

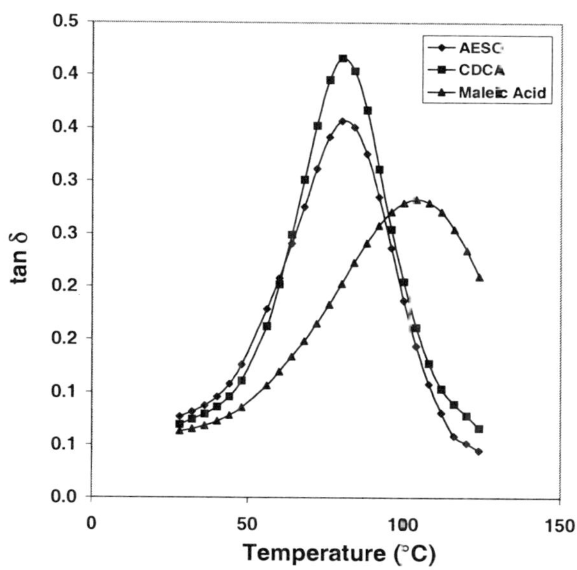

Figure 13. Damping peak (tan δ) of modified AESO resins as a function of t. Peaks in tan δ were found at 81 °C (CDCA modified) and 105 °C (maleic acid modified) compared to 79 °C for the synthesized AESO.

glass transition is rather broad due to the broad MW distribution of the SOMG maleates. The distribution of SOMG monomaleates, monoglyceride bismaleates, diglyceride monomaleates, and glycerol tris maleates has been confirmed by mass spectral analysis, as reported in (*63*). Tensile tests performed on the coploymers of SOMG maleates with styrene showed a tensile strength of 29.36 MPa and a tensile modulus of 0.84 GPa as calculated from the force displacement graph.

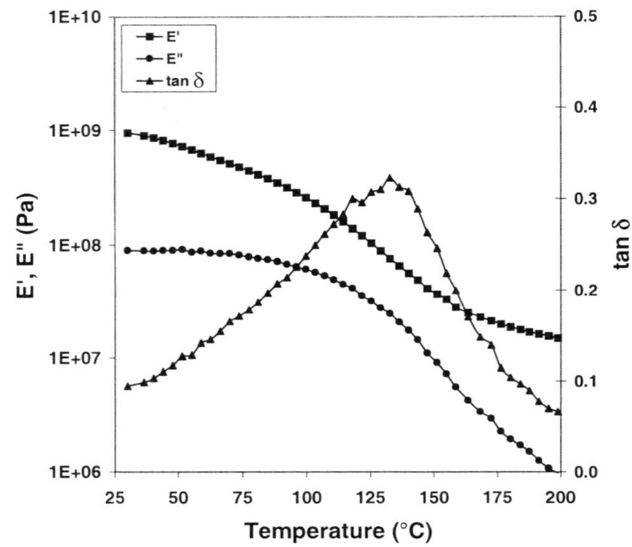

Figure 14. Dynamic mechanical behavior for SOMG/MA polymer.

SOMG/NPG Maleates (SOMG/NPG/MA)

The DMA of the SOMG/NPG/MA polymer showed a tan δ peak at about 145 °C and an E' of 2 GPa at room *t*. The 12 °C increase in T_g and the considerable increase in the modulus of the copolymers of SOMG/NPG maleates with styrene as compared to the SOMG maleates can be attributed to the replacement of the flexible fatty acid chains by the rigid methyl groups of NPG. The overall dynamic mechanical behavior of the SOMG/NPG/MA polymer was very similar to that of the SOMG/MA shown in Figure 14. Despite the higher T_g

and modulus, there remained a broad glass transition. The tensile strength of the SOMG/NPG/MA polymer was 15.65 MPa, while the tensile modulus was 1.49 GPa.

Maleinized pure NPG polymerized with styrene (NPG/MA) was prepared to compare its properties with the SOMG/NPG/MA polymer (*51*). DMA of the NPG/MA showed a tan δ peak at about 103 °C and an E' of about 2.27 GPa at 35 °C. The high T_g observed for the SOMG/NPG/MA system (~145 °C) is attributed to a synergetic effect of both the NPG and SOMG together, because the T_g observed for the NPG/MA system (~103 °C) is much lower. This is probably due to the incorporation of the fatty acid unsaturation into the polymer in the SOMG/NPG/MA system. The comparatively higher E' observed for the NPG maleates, however, explains the increase in the E' of SOMG/NPG/MA as compared to SOMG/MA. The decrease in tensile strength of SOMG/NPG/MA as compared to SOMG/MA may be attributed to a broader MW distribution of this system as compared to the SOMG maleates.

SOMG/BPA Maleates (SOMG/BPA/MA)

The DMA of this polymer showed a tan δ peak at about 131 °C and an E' of 1.34 GPa at 35°C. The introduction of the rigid benzene ring on the polymer backbone considerably increased the modulus of the final polymer as compared to the SOMG maleates. The T_g of this polymer, however, was similar to that of the SOMG maleates (133 °C). This has been attributed to a lower yield in the maleinization of the BPA, as determined from ^{1}H NMR data (*51*). Like the SOMG/NPG/MA polymer, SOMG/BPA/MA displayed the characteristic gradual glass transition shown in Figure 14.

HO/MA Dynamic Mechanical Polymer Properties

The dynamic mechanical properties of the HO/MA polymers were better than those of the AESO polymers. Little variation was seen between the polymers made from different oils. At room *t*, E' for all of the oils was between 1.45 and 1.55 Gpa and showed no dependence on saturation level. A maximum in tan δ was exhibited at 107 to 116 °C, which are all substantially higher than the AESO base resin. These properties are fairly close to those shown by conventional petroleum-based polymers. However, the distinctive triglyceride behavior still exists, in that the glass transitions are extremely broad and that even at room *t*, the materials are not completely in a glassy state. Again this is probably due to the saturated fatty acids of the triglycerides that act as a plasticizer.

Although the extent of maleinization was approximately the same from oil to oil, it is possible to see how the slight differences affect the T_g. In Figure 15, the tan δ peak *t* is plotted as a function of maleate functionality. Within this range, the behavior is linear, suggesting that if higher levels of functionalization can be reached, the properties should improve accordingly (*64*). However, it is expected that past a certain extent of maleate functionality, the tan δ peak *t* dependence will plateau. The limits of this behavior is currently being tested.

It was previously stated that the broadness in the glass transition may be inherent to all triglyceride-based polymers. However, work with genetically engineered oil and synthetic oil has shown that it is possible to reduce this characteristic. The genetically engineered high oleic soybean oil has an average functionality of 3 double bonds/triglyceride. The maleinized form of this oil had a functionality of 2 maleates/triglyceride. The properties of polymers from this material were compared to polymers from triolein oil, which is monodisperse, consisting only of oleic fatty acid esters (18 carbons long and one double bond). The maleinized triolein oil had a functionality of 2.1 maleates/triglyceride. Thus the only difference between the two oils is the fatty acid distribution of the high oleic oil vs the monodisperse triolein oil.

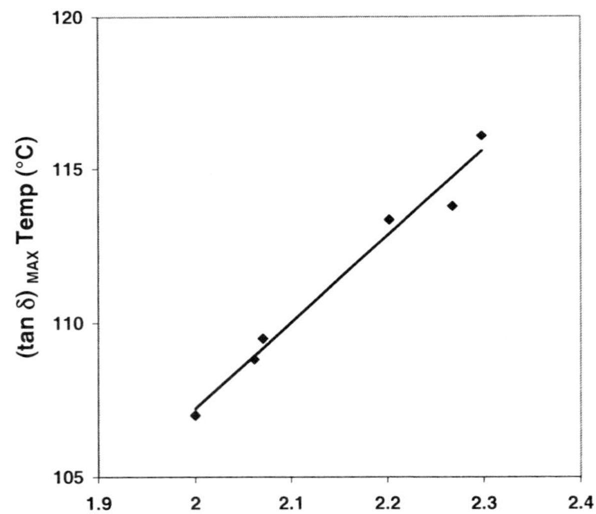

Figure 15. Peak in tan δ as a function of maleate functionality.

The dynamic mechanical properties of polymers made from these oils are shown in Figure 16. The T_g of these two polymers does not differ much, judging from either their tan δ peak or the inflection in E', but the broadness of the transitions does differ. It is apparent that the triolein polymer has a sharper E' transition from the glassy to the rubbery region. This is evident also in the tan δ peak, which has a higher peak height. The transition is not yet as sharp as petroleum-based polymers. This is probably caused by the triolein monomer only having a functionality of only two maleates per triglyceride. Consequently, there is still a plasticizer effect present, but this effect may be reduced by controlling the reaction conditions to reach higher conversions.

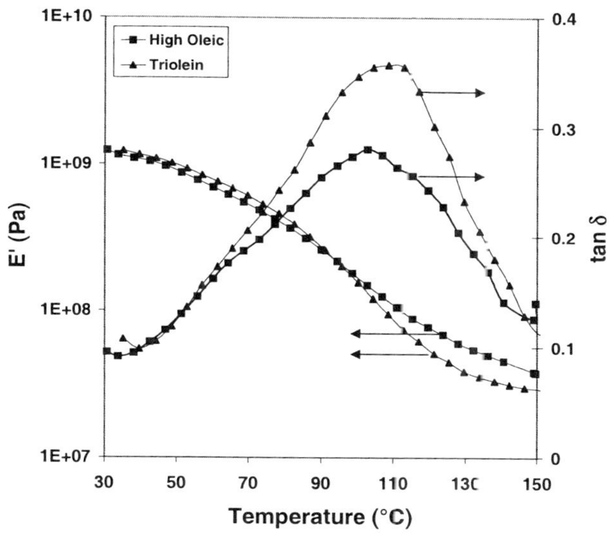

Figure 16. Dynamic mechanical properties of polymers made from maleinized hydroxylated high oleic oil and triolein oil. The monodisperse triolein displays a sharper transition from the glassy region to the rubbery region.

Conclusions

Triglyceride oils are an abundant natural resource that has yet to be fully exploited as a source for polymers and composites. The different chemical functionalities allow the triglyceride to be converted to several promising monomers. When blended with comonomers, these monomers form polymers with a wide range of physical properties. They exhibit moduli in the 1 to 2 GPa range and have T_g ranging from 70 to 120 °C, depending on the particular monomer and the resin composition. DMA shows that the transition from glassy to rubbery behavior is extremely broad for these polymers as a result of the triglyceride molecules acting both as cross-linkers and as plasticizers in the system. Saturated fatty acid chains are unable to attach to the polymer network, causing relaxations to occur in the network. However, this transition can be sharpened by reducing the saturation content, as demonstrated with the genetically engineered oil and pure triolein oil.

This area of research sets a foundation from which completely new materials can be produced with novel properties. Work is still being continued to optimize the properties of these materials and understand the fundamental issues that affect them. We have recently used computer simulation to optimize the choice of the fatty acid distribution function and determine the resulting architecture and mechanical properties for particular chemical pathways shown in Figure 1. This significantly reduces the number of chemical trials for a system with a large number of degrees of freedom and suggests the optimal oil most suited to a particular type of resin. In this manner, more renewable resources can be used to meet the material demands of many industries.

Acknowledgments

This work was supported by the National Science Foundation, the U.S. Department of Energy, the United Soybean Board, the Maryland Soybean Board, the Delaware Transportation Institute, DuPont, John Deere & Company, Hercules Inc., and Cara Plastics, Inc.

References

1. Wool, R. P.; Kusefoglu, S. H.; Palmese, G. R.;.Zhao; R; Khot, S. N.; U.S. Patent 6,121,398, 2000.
2. Wool, R. P.; *Chemtech,* **1999,** *29,* 44.
3. Liu, K. *Soybeans: Chemistry, Technology, and Utilization.* Chapman and Hall: New York, 1997.
4. Gunstone, F. *Fatty Acid and Lipid Chemistry.* Blackie Academic and Professional: New York, 1996.
5. Cunningham A.; Yapp, A. U.S. Patent 3,827,993, 1974.
6. Bussell, G. W. U.S. Patent 3,855,163, 1974.
7. Hodakowski, L.E.; Osborn, C.L.; Harris, E.B. U.S. Patent 4,119,640, 1975.
8. Trecker, D. J.; Borden, G. W.; Smith, O. W. U.S. Patent 3,979,270, 1976.
9. Trecker, D. J.; Borden, G. W.; Smith, O. W. U.S. Patent 3,931,075, 1976.
10. Salunkhe, D. K.; Chavan, J. K.; Adsule, R. N.; Kadam, S. S. *World Oilseeds: Chemistry, Technology, and Utilization.* Van Nostrand Reinhold: New York, 1992.
11. Force, C. G.; Starr, F. S. U.S. Patent 4,740,367, 1988.

12. Barrett, L. W.; Sperling, L. H.; Murphy, C. J. *J. Amer. Oil Chem. Soc.* **1993,** *70,* 523.
13. Qureshi, S.; Manson, J. A.; Sperling, L. H.; Murphy, C. J. In *Polymer Applications of Renewable-Resource Materials*; Carraher, C. E., Sperling, L. H., Eds.; Plenum Press: New York, 1983.
14. Devia, N.; Manson, J. A.; Sperling, L. H.; Conde, A. *Polym. Eng. & Sci.* **1979,** *19,* 878.
15. Devia, N.; Manson, J. A.; Sperling, L. H.; Conde, A. *Polym. Eng. & Sci.* **1979,** *19,* 869.
16. Devia, N.; Manson, J. A.; Sperling, L. H.; Conde, A. *Macromolecules* **1979,** *12,* 360.
17. Sperling, L. H.; Carraher, C. E.; Qureshi, S. P.; Manson, J. A.; Barrett, L. W. In *Polymers from Biotechnology*, Gebelein, C. G. Ed., Plenum Press: New York, 1991.
18. Sperling, L. H.; Manson, J. A.; Linne, M. A. *J. Polym. Mater.* **1984,** *1,* 54.
19. Sperling, L. H.; Manson, J. A. *J. Amer. Oil Chem. Soc.* **1983,** *60,* 1887.
20. Fernandez, A. M.; Murphy, C. J.; DeCosta, M. T.; Manson, J. A.; Sperling, L. H. In *Polymer Applications of Renewable-Resource Materials,* Carraher C. E.; Sperling, L. H., Eds.; Plenum Press: New York, 1983.
21. Sperling, L. H.; Manson, J. A.; Qureshi, S. A.; Fernandez, A. M. *Ind. Eng. Chem.* **1981,** *20,* 163.
22. Yenwo, G. M.; Manson, J. A.; Pulido, J.; Sperling, L. H.; Conde, A.; Devia, N. *J. Appl. Polym. Sci.* **1977,** *21,* 1531.
23. Frischinger; I.; Dirlikov, S. Polymer Comm. **1991,** *32,* 536.
24. Frischinger; I.; Dirlikov, S. In *Interpenetrating Polymer Networks*, Advances in Chemistry Series; Sperling, L.H., Kempner, D., Utracki, L., Eds., American Chemical Society: Washington, DC, 1994.
25. Rosch, J.; Mulhaupt, R. *Polymer Bull.* **1993,** *31,* 679.
26. Meffert, A.; Kluth, H.; Denmark Patent 4,886,893, 1989.
27. Rangarajan, B.; Havey, A.; Grulke, E. A.; Culnan, P. D. *J. Amer. Oil Chem. Soc.* **1995,** *72,* 1161.
28. Zaher, F. A.; El-Malla, M. H.; El-Hefnawy, M. M. *J. Amer. Oil Chem. Soc.* **1989,** *66,* 698.
29. Friedman, A.; Polovsky, S. B.; Pavlichko, J. P.; Moral, L. S.; U.S. Patent 5,576,027, 1996.
30. Swern, D.; Billen, G. N.; Findley, T. W.; Scanlan, J. T. *J. Amer. Chem. Soc.* **1945,** 67, 786.
31. Sonntag, N. O. V. *J. Amer. Oil Chem. Soc.* **1982,** *59,* 795.
32. Solomon,D.H *The Chemistry of Organic Film Formers*. Wiley: New York, 1967.
33. Can, E. M. S. Thesis, Bogazici University, Turkey, 1999.
34. Bailey, A. E. In *Bailey's Industrial Oil and Fat Products.* Swern, D., Ed., Wiley, New York, 1985.

35. Hellsten, M.; Harwigsson, I.; Brink, C. U.S. Patent 5,911,236, 1999.
36. Cain, F. W.; Kuin, A. J.; Cynthia, P. A.; Quinlan, P. T. U.S. Patent 5,912,042, 1995.
37. Eckwert, K.; Jeromin, L.; Meffert, A.; Peukert, E.; Gutsche, B. U.S. Patent 4,647,678, 1987.
38. Khot, S. N. M. S. Thesis, University of Delaware, 1998.
39. Wypych, J. *Polyvinyl Chloride Stabilization*. Elsevier: Amsterdam, 1986.
40. Sears, J. K.; Darby, J. R. *The Technology of Plasticizers*. Wiley & Sons: New York, 1982.
41. Carlson, K. D.; Chang, S. P. *J. Amer. Oil Chem. Soc.* **1985,** *62,* 934.
42. Raghavachar, R.; Letasi, R. J.; Kola, P. V.; Chen, Z.; Massingill, J. L. *J. Amer. Oil Chem. Soc,* **1999,** *76,* 511.
43. Pashley, R. M.; Senden, T. J.; Morris, R. A.; Guthrie, J. T.; He, W. D. U.S. Patent 5,360,880, 1994.
44. Likavec, W. R.; Bradley, C. R. U.S. Patent 5,866,628, 1999.
45. Bordon, G. W.; Smith, O. W.; Trecker, D. J. U.S. Patent 4,025,477, 1974.
46. La Scala, J. J.; Bunker, S. P.; Wool, R. P., manuscript in preparation.
47. S. P. Bunker, M.S. Thesis, University of Delaware, 2000.
48. Chu, T.-J.; Niou, D.-Y. *J. Chin. Inst. Chem. Eng.* **1989,** *20,* 1.
49. Betts, A. T. U.S. Patent 3,867,354, 1975.
50. Mitch, E. L.; Kaplan, S. L. In *Proceedings 33rd Annual SPE Technical Conference,* Atlanta, GA, 1975.
51. Can, E.; Kusefoglu, S.; Wool, R. P., manuscript in preparation.
52. Gardner H. C.; Cotter, R. J. European Patent 20,945, 1981.
53. Thomas, P.; Mayer, J. U.S. Patent 3,784,586, 1974.
54. Lee, S. H.; Park, T. W.; Lee, S. O.; *Polymer (Korea),* **1999,** *23,* 493.
55. Shione, H.; Yamada, J. Japan Patent 11,147,222, 1999.
56. Hasegawa, H. Japan Patent 11,240,014, 1999.
57. Johnson, L. K.; Sade, W. T. *J. Coat. Tech.* **1993,** *65,* 19.
58. Solomons, T. W. G. *Organic Chemistry*; Wiley: New York, 1992.
59. J. La Scala and R. P. Wool, *Polymer,* in preparation.
60. Flory,P.J. *Principles ofPolymer Chemistry*.Cornell University:Ithaca,NY 1975.
61. Nielsen, L. E.; Landel, R. F. *Mechanical Properties of Polymers and Composites*, Marcel Dekker: New York, 1994.
62. Wool, R. P.; Khot, S. N. In *Proceedings ACUN-2*, Sydney, Australia, 2000.
63. Kusefoglu, S. H.; Wool, R. P.; Can, E. *J. Appl. Polym. Sci.,* **2001**,*81*,69.
64. S. N. Khot, La Scala, J.J., Can E., Morye, S.S., Williams, G.I., Palmese, G., Kusefuglu, S.H., and Wool, R.P., *J Appl. Polym. Sci.*, **2001**, 82, 703.

Chapter 14

Affordable Composites and Plastics from Renewable Resources: Part II: Manufacture of Composites

Richard P. Wool, Shrikant N. Khot, John J. LaScala, Shana P. Bunker, Jue Lu, Wim Thielemans, Erde Can, Shantaram S. Morye, and George I. Williams

Department of Chemical Engineering, ACRES Program, Center for Composite Materials, University of Delaware, Newark, DE 19716

When bio-based resins derived from natural oils (described in Part I in this book) are combined with natural fibers, glass fibers, and lignin, new low-cost composites are produced that are economical in many high-volume applications. These composites are used in agricultural equipment, automotive sheet molding compounds, civil and rail infrastructures, marine applications, housing and the construction industry. Examples are given for the synthesis, manufacture, and properties of plant-based resins and glass, flax, lignin, and hemp composites, including Sheet Molding Compound (SMC). In addition to the highly cross-linked thermoset composite resins, Pressure Sensitive Adhesives (PSA) consisting of high molecular weight linear and branched chains derived from functionalized triglycerides are also presented.

Introduction

Triglyceride-Based Composite Materials

All of the resins presented here are suitable for use as a matrix in a composite material. Their low viscosity and method of curing make them ideal candidates for use in conventional resin transfer molding (RTM) processes. Most polymer matrix composites are made by embedding strong fibers such as carbon, aramid, glass, or natural fibers in a polymer matrix. The high strength and modulus of the embedded fibers impart strength and rigidity to the material that surpass that of the neat polymer (*1*). In recent years natural fibers have attracted attention as potential reinforcements due to the high cost of synthetic fibers. These cellulose-based fibers offer the advantages of biodegradability, low density, nonabrasive nature, and low cost.

Depending on their origin, natural fibers can be grouped into seed, bast, leaf, and fruit qualities. Bast fibers (e.g., hemp, jute, flax, ramie, kenaf) and leaf quality fibers (e.g., sisal, banana leaf) are the most commonly used in composite applications. Properties for these fibers include excellent tensile strength and modulus, high durability, low bulk density, good moldability, and recyclability. These natural fibers have an advantage over glass fibers in that they are less expensive, abundantly available from renewable resources, and have a high specific strength. While high performance carbon fibers remain superior to natural fibers in high-end applications, natural fibers have comparable properties to glass fibers in high volume applications (*2*). The properties of flax, jute, sisal, and hemp fibers are shown in Table I and are compared to the commonly used E-glass fiber (*3*). Notably, flax fiber has a modulus higher than that of E-glass. Flax is also less dense and so produces a lighter composite with good mechanical properties.

Table I. Properties of Natural and E-Glass Fibers (*3*)

	Density (kg/m^3)	*Tensile Modulus (GPa)*	*Tensile Strength (GPa)*
Flax	1500	100	1.1
Jute	1450	2.5 to 13	0.46 to 0.53
Sisal	1450	9.4 to 15.8	0.57 to 0.64
Hemp	1480	—	0.69
E-glass	2540	76	1.5

Numerous studies on the properties of natural fiber composites using jute (*4–8*), banana (*9*), agave (*9*), hemp (*9, 10*), flax (*10–12*), bamboo (*13*), pineapple (*14*), and rubber wood (*15*) have appeared in the literature. For certain applications natural fiber composites are not sufficient because of the low strength of these fibers. However, combining natural fibers with stronger synthetic fibers like glass could offer an optimum balance between performance and cost. These "hybrid" composites, which use two different types of fiber, have been examined in such forms as jute/glass hybrids with epoxy and polyester matrix materials (*16, 17*).

In previous works, natural fibers were combined with petroleum-derived matrix resins. The resins presented here offer the unique potential of combining natural fibers with resins based on natural renewable resources. Here we present the properties of glass reinforced composite materials made from AESO resin as well as all-natural fiber composite materials reinforced by flax and hemp fibers (*18*). We also review the properties of hybrid composites made from AESO-based resins reinforced with flax and glass fibers (*19*).

Experimental

Manufacture of Glass Fiber Reinforced Composites

The properties of glass fiber reinforced composites made from AESO and HSO/MA polymers (*20*) were examined by mixing 1500 g of AESO with 750 g styrene, 113 g divinyl benzene, 18 g cobalt napthalate activator, and 68 g Trigonox 239 free radical initiator. The resin was infused into a glass fiber (50% volume fraction) preform by Seeman's composite resin injection molding process. The composite was cured for 12 h at room temperature (*t*) and post-cured at 150 °C for 2 h. The properties were compared with those of a commercial high performance vinyl ester resin containing the same fiber volume fraction and manufactured under the same conditions.

The HSO/MA composites were made using a resin composed of 100 g HSO/MA monomer synthesized in the manner previously mentioned, 45 g styrene, 5 g divinyl benzene, and 2.25 g 2,5-dimethyl-2,5-di-(2-ethylhexanoyl-peroxy) hexane as an initiator. The mixture was injected into a resin transfer mold containing two mats of glass fiber. The mold was heated at 65 °C for 1.5 h and was post-cured at 120 °C for 1 h. Properties were compared with those of the vinyl ester resin Dow Derakane 411C50 prepared under the same conditions.

The tensile, flexural, and compressive properties for both composites were measured per ASTM D 3039, ASTM D 790, and ASTM D 3410, respectively.

Manufacture of Flax and Hemp Composites

Composites were manufactured using Durafiber Grade 2 flax fiber and hemp fibers within the AESO-based polymer. The hemp was obtained in the form of a non-woven mat with an oriented fiber direction. The method used to manufacture the flax and hemp composites is detailed elsewhere (*18, 21*).

The resin used in these all-natural composites was prepared by blending AESO, styrene, and divinylbenzene in the ratio 100:45:5 by weight (wt). To initiate free-radical polymerization, 1.5% of the total resin wt of 2,5-dimethyl-2,5-di-(2-ethylhexanoylperoxy)hexane was added. The composite was manufactured using a RTM process. For the flax composites, flax fibers were uniformly arranged in the mold in a random mat. Hemp composites were made with two layers of the oriented hemp mats placed with their orientation perpendicular to each other. Resin was injected into the mold, and the composite was cured at 90 °C for 1 h, followed by a post-cure at 110 °C for 1 h. The tensile and flexural properties were measured per ASTM D 3039 and ASTM D 790, respectively.

Manufacture of Glass-Flax Hybrid Composites

Hybrid composites were manufactured symmetrically and asymmetrically using E-Glass woven fiber and Durafiber Grade 2 flax fibers in an AESO-based polymer (*19*). In the symmetric composites, a layer of flax fiber was sandwiched between two layers of glass fiber. The asymmetric composites were produced by uniformly arranging the flax fibers at the bottom of the mold in a random mat and then placing two layers of woven glass fabric on top.

The resin for these studies was prepared by mixing AESO with 50 parts per hundred by wt (phr) styrene, 4.5 phr Trigonox 239A free radical initiator, and 1.2 phr cobalt naphthalate activator. After injecting the resin into the mold, the composite was cured overnight at 2.1 MPa and room *t*. The composite was post-cured at 110 °C for 2 h, after which the composite panels had fiber wt fractions in the range of 0.31 to 0.40. Composites with glass/flax ratios of 100/0, 80/20, 60/40, 40/60 and 0/100 were made. The tensile, compression, and flexural properties were measured per ASTM D 3039, ASTM D 3410, and ASTM D 790, respectively. Impact tests were conducted per ASTM D 3763.

Results and Discussion

AESO and HSO/MA Glass Fiber Composites

The properties of the glass fiber reinforced AESO composites are shown in Table II (*20*). The tensile strength, tensile modulus, and compressive modulus all are similar to the properties of the vinyl ester resin. The only apparent shortcoming is in the compressive strength of the AESO resin, which can be attributed to the lower strength of the AESO neat polymer. However, according to DMA, the AESO composite still displayed a T_g close to that of the neat polymer at approximately 80 °C. This is much lower than the T_g of the vinyl ester polymer, which was found to be about 128 °C.

Table II. Properties of Glass Fiber Reinforced AESO-Based Polymer and Dow PC100 Vinyl Ester Polymer

	Testing Direction	*Tensile Strength (MPa)*	*Tensile Modulus (GPa)*	*Compressive Strength (MPa)*	*Compressive Modulus (GPa)*
AESO	0°	463.2	24.8	302.6	24.8
Dow PC100	0°	458.4	23.8	420.5	23.4
AESO	90°	321.9	20.7	130.6	20.7
Dow PC100	90°	324.0	17.6	339.1	17.9

The HSO/MA composite was even more successful at replicating the properties of a vinyl ester composite, as shown in Table III. The flexural modulus and compressive strength for the HSO/MA composite were of same magnitude as the vinyl ester composite, while the flexural strength was slightly lower. Additionally, T_g for the HSO/MA composite was approximately 128 °C, which equals that found for the vinyl ester composite.

These results indicate that although the properties of the neat soy oil-based polymers are less than those of the vinyl ester polymers, the composite material properties are very similar. In tensile deformation, the fiber reinforcement is able to support the majority of the load leading to an acceptable modulus and strength. The area that needs improvement is compression deformation, where the polymer bears the majority of the stress.

Table III. Properties of HSO/MA-Based Polymer and Dow 411C50 Vinyl Ester Polymer

	Flexural Modulus (GPa)	*Flexural Strength (MPa)*	*Compressive Strength (MPa)*
HSO/MA	34.5	669	200
Dow DK 411C50	35.8	813	290

Flax Composites

Figure 1 shows tensile and flexural strength as a function of fiber content for the Durafiber Grade-2 flax composites. The tensile strength of the AESO/flax fiber composite was a maximum of 30 MPa at 34% fiber content, which is comparable to the tensile strength of AESO neat resin (~30 MPa). The flexural strength and flexural moduli of these materials all showed a maximum at 34% fiber content, while the tensile moduli increased with fiber content, Figure 2. This optimization phenomenon has been explained previously in terms of increasing fiber-fiber interactions as fiber content increases (*2*). This reduces the level of fiber-matrix interaction, thereby weakening the composite. Percolation theory has also been used to explain this effect (*22*).

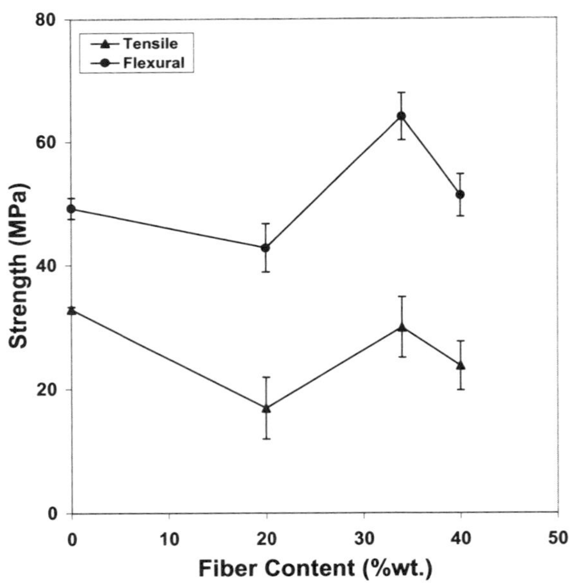

Figure 1. Strength dependence on composition for flax (Durafiber Grade-2) reinforced AESO polymer.

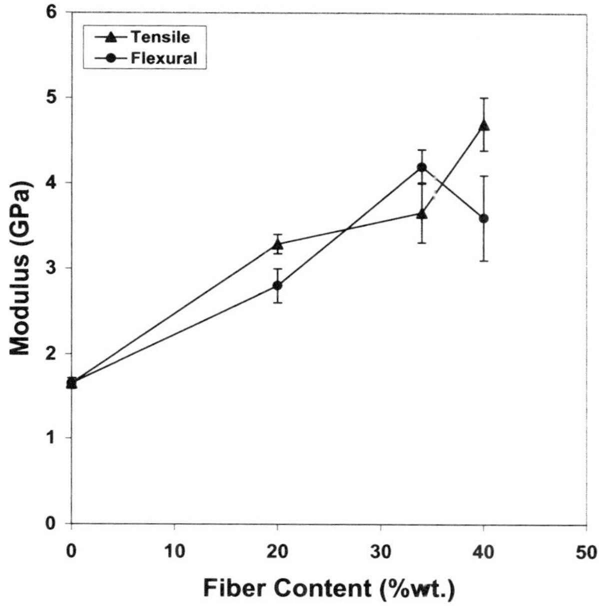

Figure 2. Modulus dependence on composition for flax (Durafiber Grade-2) reinforced AESO polymer.

Hemp Composites

Composites made of 20 wt % hemp fiber were found to display a tensile strength and modulus of 35 MPa and 4.4 GPa, respectively. The flexural properties of the composites were found to be anisotropic. Samples tested with the side of the composite in contact with the upper surface of the mold during the curing process display a yield strength of 35.7 ± 5.9 MPa and a modulus of 2.6 ± 0.2 GPa. Samples tested with the side in contact with the lower surface of the mold during cure have a yield strength of 51.3 ± 2.7 MPa and a modulus of 2.7 ± 0.2 GPa. Thus, the modulus does not appear to be affected by orientation, while the yield strength has a 44% increase. This increase in strength is substantial and can be attributed to the orientation of the fibers in the mat.

The mechanical properties of the all-natural composites are comparable to the properties of wood. For example, a typical hard wood has a tensile modulus of about 10 GPa, with a fracture stress of about 30 MPa when stress is exerted parallel to the fiber axis and of about 3 MPa when stress is exerted normal to the fiber axis. The considerable advantage of the all-natural composites is that the unidirectional high strength properties of wood can be obtained in all directions for the randomly oriented fiber composite. In addition, the ease of manufacturing complex shapes via normal composite liquid molding operations provides a significant cost advantage for these materials. The ACRES group is currently designing roofs for houses using these materials, which are expected to be more hurricane-resistant than current designs.

Hybrid Natural-Glass Fiber Composites

The tensile modulus, tensile strength, and compressive strength of the glass/flax hybrid composites for different glass/flax ratios and composite constructions are shown in Table IV. As expected, these properties all increase with increasing glass fiber content. The 100% flax fiber reinforced materials show a tensile strength and modulus of 26.1 ± 1.7 MPa and 1.9 ± 0.1 GPa, respectively. At the other extreme, the 100% glass fiber reinforced materials show a tensile strength and modulus of 128.8 ± 1.1 MPa and 5.2 ± 0.1 GPa, respectively.

Table IV. Properties of Symmetric Glass/Flax Hybrid Composites

Glass/Flax Ratio	*Wt Fractions Glass*	*Wt Fractions Flax*	*Tensile Modulus (GPa)*	*Tensile Strength (MPa)*	*Compression Strength (MPa)*
100/0	0.35	0.00	5.2 ± 0.1	128.8 ± 1.1	89.8 ± 3.2
80/20	0.25	0.06	3.5 ± 0.1	123.3 ± 1.2	71.6 ± 2.6
60/40	0.23	0.16	3.2 ± 0.1	109.1 ± 1.0	62.3 ± 3.1
40/60	0.16	0.24	2.9 ± 0.2	82.6 ± 1.4	33.6 ± 0.8
0/100	0.00	0.31	1.9 ± 0.1	26.1 ± 1.7	18.5 ± 2.4

As shown in Table V, the asymmetric and symmetric composites have similar tensile moduli. However, the tensile and compression strengths of the asymmetric composites were noticeably less than those of the symmetric composites. This is due to the different modes of failure exhibited by the two types of composites. The symmetric composites undergo tensile failure at the peak load, while the asymmetric composites fail by shear delamination at the glass/flax interface because of the different tensile moduli of the two fiber types.

The flexural properties of the glass/flax hybrid composites are shown in Table VI. The flexural modulus and strength for the glass fiber composite are much higher than those for the flax fiber composite due to the higher modulus and strength of glass fibers. The 100% flax-reinforced composites display a flexural strength and modulus of 61.0 ± 3.4 MPa and 3.8 ± 0.2 GPa, respectively. The 100% glass fiber reinforced composites have a flexural strength and modulus of 205.5 ± 4.5 MPa and 9.0 ± 0.2 GPa, respectively. Additionally, there is an obvious anisotropy in the behavior of the asymmetric composites depending on the surface that bears the load. The maximum flexural strengths occurred when the flax surface bore the exerted load or impact. In such an orientation, the glass fibers bear a tensile load from the bending of the sample.

Table V. Properties of Asymmetric Glass/Flax Hybrid Composites

Glass/Flax Ratio	*Wt Fractions Glass*	*Flax*	*Tensile Modulus (GPa)*	*Tensile Strength (MPa)*	*Compression Strength (MPa)*
80/20	0.25	0.06	3.4 ± 0.1	111.7 ± 2.1	65.3 ± 4.8
60/40	0.24	0.16	3.1 ± 0.1	90.6 ± 2.4	46.2 ± 0.6
40/60	0.16	0.25	2.7 ± 0.3	68.9 ± 2.1	30.1 ± 2.2

The impact energy of the hybrid composites ranged from 13.3 ± 0.3 J to 28.7 ± 1.2 J. The maximum impact energy absorbed (28.7 ± 1.2 J) was shown by the asymmetric 40/60 glass/flax ratio composite when the flax surface was the load-bearing face. The energy absorption by the symmetric hybrid composites seems to be only marginally higher than that of the 100% glass fiber composite, a difference made even more insignificant when considering the standard deviations.

Sheet Molding Compound

Sheet Molding Compound (SMC) is the equivalent of metal sheet stamping in the auto industry and consists of a B-Staged partially hardened glass/filler reinforced resin with the consistency of leather, which is then fully cured with pressure and *t* in the mold. SMC is a less expensive manufacturing method than RTM or Hand-Lay-up for high volume, and high production rates of large parts, such as used in agricultural equipment. The ACRES group participated in the design of new soy-based SMC materials for John Deere & Company agricultural equipment. Prototype soy-based composite parts for Round Hay Balers, tractors and harvesters were made by RTM methods. However, mass production of soy-based composites in agricultural equipment must be done by SMC and not RTM of liquid molding resins. The SMC materials will have a range of price, density, recylability and biodegradability. A typical SMC part will have the approximate composition given in Table VII.

An important part of SMC manufacture is the development of the B-Staged thickened resin. This requires the addition of a suitable number of carboxyl groups to the soy resin monomers, either by co-monomer addition or by direct carboxylation of the soy resins. We examined several chemical pathways for the optimal bio-based SMC resin formulation and successfully developed a B-Stage

Table VI. Flexural Properties and Energy Absorption on Impact of Glass/Flax Hybrid Composites

Glass/Flax Ratio	*Wt Fractions Glass*	*Wt Fractions Flax*	*Composite Construction*	*Loading/ Impact Face*	*Flexural Modulus (GPa)*	*Flexural Strength (MPa)*	*Energy Absorbed (J)*
100/0	0.35	0	---	---	9.0 ± 0.2	205.5 ± 4.5	16.5 ± 0.2
80/20	0.25	0.06	Symmetric	---	6.9 ± 0.2	130.3 ± 3.0	17.7 ± 1.9
	0.25	0.06	Asymmetric	Glass	6.3 ± 0.3	87.8 ± 3.9	13.3 ± 0.3
			Asymmetric	Flax	5.0 ± 0.1	189.0 ± 8.5	25.8 ± 1.1
60/40	0.23	0.16	Symmetric	---	6.0 ± 0.2	115.3 ± 2.5	18.0 ± 0.3
	0.24	0.16	Asymmetric	Glass	4.0 ± 0.3	80.1 ± 0.7	14.7 ± 0.3
			Asymmetric	Flax	4.7 ± 0.3	146.9 ± 5.5	27.6 ± 2.6
40/60	0.16	0.24	Symmetric	---	5.8 ± 0.5	83.3 ± 5.4	18.5 ± 0.2
	0.16	0.25	Asymmetric	Glass	3.8 ± 0.1	73.2 ± 7.5	15.1 ± 0.3
			Asymmetric	Flax	3.3 ± 0.4	111.1 ± 9.5	28.7 ± 1.2
0/100	0	0.31	---	---	3.8 ± 0.2	61.0 ± 3.4	1.4 ± 0.2

resin system that gave excellent mechanical properties when cured. The carboxyl groups complex with the divalent cations (manganese dioxide, for example), thicken the resin near a gel point at room *t*, and if done correctly in the presence of fibers, fillers, etc.), form the B-Stage composite. The B-Stage resin can then be fully cured at higher *t*'s to give the final SMC product. While the B-Stage resin is partially gelled by ionic complexation at room *t*, this complex must break down at the higher molding *t*'s and allow the SMC to conform to the mold before the final free radical cure occurs. The ionic complexation and ion clustering in the cured state also provides new mechanisms for the design of self-healing composites (*23, 24*).

Table VII. Sheet Molding Compound

Component	*Wt Percent Ranges*
Styrene	5–14%
Soy resin	10–12%
Styrene/soy ratio	0:1–1:1
Glass	25–35%
Glass fiber sizing	—
Natural fibers	0–35%
Natural fiber sizing	—
Calcium carbonate	35–45%
Low profile additive	3–4%
Initiator	1.00%
Divalent cation	0.5–1.0
Zinc stearate	1.00%
Other additives	0–10%

The new plant-oil based SMC resins can be stored for long periods by the addition of suitable inhibitors and will cure on command upon addition of suitable catalysts either at room *t* or at elevated *t*'s. When commonly used thickeners such as MgO dispersion are added to the resin system, the compound will reach molding viscosity in 2 to 4 days and will plateau at the moldable viscosity thereafter. Upon heating, the resin viscosity quickly decreases to ensure good flow in the mold. The bio-based SMC resins also show very good compatibility with natural fibers such as flax, jute, hay, hemp, and cotton.

Lignin-Modified Resins and Natural Fibers

Lignin is one of the three most abundant renewable resources on our planet along with cellulose and natural oils. Most industrial lignin is obtained as a waste product during the paper pulping process, but it can be found in all plants like rice and straw. While cellulose is used for paper production, natural oils are mainly used in the food industry, and the industrial applications of lignin are rather limited, despite its widespread availability. In 1998, about 1% of all lignin generated in paper production worldwide was isolated and sold (*25*). The remaining 99% was either burned in an energy recovery step for the pulping process or was disposed of in waste streams. The global production of lignin products was reported to be 138,500 ton/yr in 1990 (*26*). Extrapolating these 1% numbers leads to a worldwide production of more than 10 million ton/yr of available renewable raw material.

A large number of studies on lignin have focused on finding a higher value application for lignin than fuel (*25, 27*). Applications of water-insoluble kraft lignin in polymers include filler use in thermoplastics (*28*), co-reactant for thermoplastics (*29*), and co-reactant in thermosetting phenolic (*30, 31*), epoxy (*32*), and polyurethane (*27, 33*) resins. Lignin is a promising compound to be used in polymers because of its phenolic base structure (see Figure 3), which could lead to improvement of mechanical properties when incorporated in a plastic. Studies concentrating on the use of lignin as a co-monomer in unsaturated polyesters or vinylesters are limited and deal with acrylics (*34*). This is largely due to the free radical inhibiting capability of the quinonic structures in the kraft lignin molecule (*27*), and the ability to form more of these quinonic structures by resonance stabilization from aromatic hydroxyl groups, as well as the unavailability of suitable double bonds reactive towards free radical polymerization. Additionally, lignin displays incompatibility with styrene, which is commonly used as a reactive diluent in unsaturated polyester resins. More recently, Meister explored the grafting of monomers onto lignin followed by copolymerization with a broad range of co-monomers via free radical and ionic polymerization (*35*).

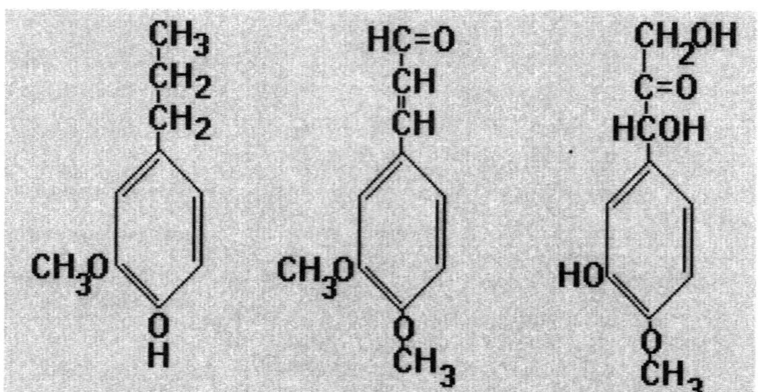

Figure 3. Structure of lignin.

Kraft lignin has a T_g of about 142 °C (*29*), which can be increased by chain extension reactions (*36*). Thus, adding lignin to existing resin systems has the potential to increase the T_g's of both petroleum-based and ACRES unsaturated polyester resins (*37*). Lignin has also been proven to increase the electrical resistance in polyolefins, making them more attractive in electrical shielding applications (*28*). Lignin in wood protects the cellulose fibers against biological attack, in addition to adhesion of the cellulose fibers. However, lignin is degradable by certain wood-rotting fungi (*38*), such as White Rot, thus making it possible to break down lignin-based materials in some environments. Using lignin in ACRES resins could result in a more controlled biodegradable resin.

The ACRES resins used in this research are derived from soybean oil. The unsaturated carbon-carbon bonds found in this oil are relatively nonreactive toward free radical polymerization, and functionalization of these double bonds is therefore needed. Four chemical pathways relevant to this research were previously described in Part I in this book.

The easiest way to introduce a new component in an existing resin is by adding the unmodified component to the system as a filler, which may change certain properties (mechanical, electrical, etc.) of the matrix resin in a positive way. In thermoplastic polyolefins, the addition of lignins showed no significant increase in mechanical properties (*28*), and therefore thermoset resins were considered in this study. The high styrene content in commercially available thermosets (about 50%), such as vinylesters and unsaturated polyesters, showed incompatibility with lignin. For this reason, only soybean oil based resins, with the ability to vary the styrene content, were used for this study.

The aliphatic and aromatic hydroxyl functional groups present in lignin structures (Figure 3) also provide a possibility for an esterification reaction with maleic anhydride to introduce reactive maleate half-ester groups to the molecule. The double bonds of the maleate half esters are reactive for radical polymerization. Although homopolymerization of lignin-maleates using the introduced double bond functionality seems impossible due to the limited reactivity of lignin with maleic anhydride, the introduction of this product to other similar resin systems containing maleate functional groups may provide some improvements in the mechanical properties of the resulting polymers by connecting this large molecule to the resin through the maleates. In this study we explored the effect of the addition of the lignin maleates to the SOMG/MA. When copolymerized with styrene, SOMG/MA produce rigid thermoset polymers. Lignin, which is a natural product and a large polyol molecule, has a potential to be used as a co-polymer in the SOMG/MA resin system.

To make lignin more compatible with commercial vinylesters and unsaturated polyesters with a high styrene content, the aliphatic and aromatic hydroxyl groups of lignin can also be reacted with an epoxy such as ESO, as is done to introduce lignin into epoxy resins (*32*). The remaining unreacted epoxy groups can later be reacted with acrylic or maleic acid, for example, to provide the double bonds to anchor the lignin molecules in the resin.

The mechanical properties of polymer composites depend on the properties of the fiber, the properties of the matrix, and the interfacial strength, among other factors. A strong interface is required for efficient transfer of stress from the matrix to the reinforcing fibers. The fibers are therefore treated with sizing or bonding agents to improve the interfacial strength. This issue is especially important in natural fiber composites. Natural fibers, being cellulosic in nature, have surfaces that are highly hydrophilic. This results in poor bonding with polymeric matrices that are organic in nature. A number of treatments have been used in the past with varying success to impart bonding between natural fibers and polymeric matrices (*4–6, 13–15*). In the current study, experiments were carried out to explore the application of lignin as a surface treatment agent for natural fibers. This concept is derived from nature, because wood can be considered a natural composite in which cellulose fibers are glued together in a matrix of lignin. The hydrogen bonds provide the necessary strength and stiffness to wood. Lignin therefore has the potential to be used as a low-cost surface treatment agent for composites of natural fibers with petroleum based resins like vinyl ester, epoxy, and unsaturated polyester.

Exploratory work by Thielemans (*39*) examined novel applications, such as filler use and co-monomer, for lignin in thermosetting unsaturated polyesters and vinylesters. The solubility of different lignins (pine kraft, hardwood, ethoxylated, and maleinated lignin) was determined in different resin systems (AESO, Hydroxylated Soybean Oil (HSO), Soy Oil Monoglyceride (SOMG), and a commercial vinylester) to give an idea of compatibility of lignin with these resin systems. The results are shown in Table VIII.

The lignin particles that did not form a sediment either dispersed or dissolved in the resin. The dispersion can be attributed to resin viscosity or affinity of lignin for the resin. These methods of dispersion are only important for the smaller particles, because gravitational forces are too strong for the larger particles to be outweighed by affinity or even viscosity, in the viscosity range of the resins used here. The controlling dispersion effect, either affinity or viscosity, is shown in Table VIII. Note that only the dispersed and dissolved lignin is being accounted. There was some sedimentation of lignin in all the samples. The poor dispersion of lignin-MA is mainly due to the large particle size of the ground material.

It is notable that there is a small lignin fraction that dissolves in the low polarity vinylester resin. Lignin particles were hardly visible with the naked eye. The solubility of lignin in SOMG is much better than in vinylester, while HSO is somewhere in between. This could be seen by the darkness in color of the solution. The results in Table VIII could be anticipated from the resin's polarity. The solubility behavior matches the relative polarity of the resins. Because lignin only dissolves in highly polar solvents, complete dissolution was not obtained in any of the systems. To dissolve lignin completely in composite resins, lignin will have to be chemically modified.

Table VIII. Ruling Effect for Dispersion of Lignin in Different Resins

	Pine lignin	*Hardwood lignin*	*Sulfonated, Ethoxylated Lignin*	*Lignin-MA adduct*
AESO	viscous dispersion	viscous dispersion	viscous dispersion	viscous dispersion
HSO	affinity dispersion	no dispersion	affinity dispersion	no dispersion
SOMG	affinity dispersion	affinity dispersion	affinity dispersion	no dispersion
Vinylester	soluble	soluble	no dispersion	soluble

The use of lignin as a filler was studied. An increase in T_g was noticed, while the modulus at 20 °C decreased due to a plasticizing effect of lignin. Then, lignin was modified to improve its effect on the matrix properties by adding double bond functionality, thus making it possible to incorporate the lignin molecule in the resin through free radical polymerization. Modified lignin, by reaction with maleic anhydride (MA) and epoxidized soybean oil (ESO), was introduced in several resins and tested for its effect on solubility, T_g, and modulus. This modification improved the solubility of lignin in styrene containing resins as well as the chemical incorporation of lignin in the resin. Furthermore, lignin was used to treat the surfaces of natural hemp fibers to use lignin's natural affinity for cellulosic fibers. The idea was to cure the surface defects on the natural fibers as well as to increase the bonding strength between resin and fiber. An optimum improvement was noticed depending on the amount of lignin covering the fibers, as shown in Table IX.

When the hemp fibers are treated with a 10% solution of lignin, the resultant composites contain 6.3% lignin by wt of the composite. This composite has lower tensile and flexural strengths compared with the untreated hemp fiber composites. The tensile and flexural moduli are, however, largely unchanged. Hemp fibers were also treated with a lignin solution of a lower concentration (1%). These composites were found to have 0.2% lignin by wt of the composite and showed improved properties; the tensile modulus increased by 10.0%, the flexural modulus increased by 22.5% and the flexural strength increased by 7%. Thus, lignin potentially can be a useful sizing agent for natural fiber composites.

Recent work with lignin by Thieleman and Wool has successfully determined methods of solubilizing the lignin in non-polar resins. This allows

Table IX. Mechanical Properties of Hemp Composites with Lignin Surface Treatment

Composite	W_{hemp} *(%)*	W_{lignin} *(%)*	W_{VE} *(%)*	*Density* *(kg/m³)*	*Tensile Modulus (GPa)*	*Tensile Strength (MPa)*	*Flexural Modulus (GPa)*	*Flexural Strength (MPa)*
Hemp/VE	20.0	0.0	80.0	1118.7 ± 2.3	3.9 ± 0.1	44.9* ±0.9	4.0 ± 0.1	80.5 ± 2.9
Lignin-hemp/VE	19.1	6.3	74.6	1123.5 ± 2.4	3.7 ±0.1	32.7* ± 0.7	3.9 ± 0.1	53.4 ± 1.6
Lignin-hemp/VE	19.9	0.2	79.9	1150.8 ± 2.0	4.3 ±0.1	37.9* ± 0.4	4.9 ± 0.2	86.2 ± 2.5

NOTE: * indicates that failure happened near grips instead of in gauge length region.

lignin to be used as a toughening agent, interfacial agent and chemical additive in the conventional or the bio-based composite resins described in Part I.

Pressure Sensitive Adhesives

The demand for PSAs has been increasing over the last several years, and the current world market is estimated at about 14 billion pounds. Applications include transparent tape, duct tape, masking tape, nametags and a variety of products that are typically disposable. Recently, Bunker and Wool (*40*) investigated the synthesis of these adhesives from plant oils. Plant oil derivatives, such as fatty acid methyl esters, provide an ecological advantage over current PSA technology, which is largely based on petrochemical derived acrylics. Both triglycerides as well as individual fatty acids can be chemically modified to participate in free radical polymerization reactions and generate linear, branched, and lightly cross-linked polymers. Given the floppiness of triglyceride molecules, they are perhaps more suited to forming tacky soft materials for PSA than the high modulus hard composite resin materials, as described in the previous sections. The chemical modifications include placing an C=C substituted carboxylic acid, such as acrylic acid, onto the unsaturation present in the starting material, as described in Part I in this book.

Because of the variable unsaturation in plant oils, there are many possible chemical routes in which different chemical functionalities can be placed on the oil substrate. This chemical flexibility will allow us to design substrate-specific adhesives or a general adhesive. In this work, we focus on using acrylate functionality, where the starting material was a high purity oleic methyl ester. This material is a long aliphatic chain with 18 carbons and exactly one double bond ($C_{19}O_2H_{36}$). The single double bond allows us to easily synthesize linear polymers. To place an acrylate group on the methyl ester backbone, we first epoxidized the double bond. This is performed through a standard reaction using hydrogen peroxide and formic acid. The epoxy group is then susceptible to a nucleophilic substitution by acrylic acid.

The polymer is formed using emulsion polymerization. Due to the ultra low solubility of the starting monomer ($<10^{-7}$ M) (*41*), acrylic acid is used sparingly in small quantities as a co-monomer to seed the polymerization reaction. The T_g of the material was analyzed using dynamic scanning calorimetry and found to be about –30 °C, which is typical for a commercial acrylic-base PSA. The T_g of these materials can be raised or lowered by copolymerization with rigid or flexible groups, respectively.

The M_W of the emulsion polymerized acrylated oleic methyl ester (AOME), analyzed by gel permeation chromatography, was on the order of 10^6 to 10^7 g/mol. Because the critical entanglement molecular wt (M_c) is on the order of $2x10^5$ for this material, M_W's of about 10^6 are required to obtain good tack properties. The peel energy requirement to debond a PSA tape from a substrate ranges from 1 to 10 lb_f/in (pli), and 180° peel tests have shown that the peel energy of these polymers ranges from 1.5 to 3 pli. To increase peel energy, it is common to add a slight degree of cross-linking to the polymer system. By manipulating the gel to non-gel ratio, a higher performance adhesive can be designed. We can vary the cross-link density to suit most adhesive debonding requirements by the addition of acrylated di-and triglycerides. The shear time to failure of these materials was increased from 15 min to 25 h by the addition of a small amount of these di- and triglycerides. The tack and creep of the new bio-based PSA resins can be controlled both by designing suitable molecular architectures (linear, branched, cross-linked) and by formulation with tackifier additive. The initial properties of the new bio-based PSA are comparable to commercial PSA and show considerable promise for the future.

The low-cost production of mono-functionalized fatty acids required to make linear polymer chains can be considerably enhanced with genetic engineering, such as with DuPont's new high-oleic oils, where the oleic content is about 80%. It is apparent that for each resin there exists an ideal fatty acid distribution function that could be grown specifically to suit that resin. Thus, olive oil is more suited to PSA, while linseed oil is more suited to composite resins. Combinations of the two could be useful in making elastomers with controlled cross-link density.

Conclusions

The resins derived from plant oils are suitable for use in many molding processes to produce composite materials. At low glass fiber content (35 wt %), composites produced from AESO by RTM displayed a tensile modulus of 5.2 GPa and a flexural modulus of 9 GPa. They also exhibit a tensile strength of 129 MPa and flexural strength of 206 MPa. At higher fiber contents (50 wt %) composites produced from AESO displayed tensile and compression moduli of 24.8 GPa each. The tensile and compressive strengths were found to be 463.2 and 302.6 MPa, respectively. Besides glass fibers, natural fibers such as flax and hemp can be used in the composite materials. Hemp composites of 20% fiber content display a tensile strength of 35 MPa and a tensile modulus of 4.4 GPa. The flexural modulus was approximately 2.6 GPa and flexural strength in the range of 35.7 to 51.3 MPa, depending on the test conditions. The flax composite materials, also, have tensile and flexural strengths in the ranges of 20 to 30 MPa and 45 to 65 MPa, respectively. The properties exhibited by both the natural and synthetic fiber reinforced composites can be combined through the production of "hybrid" composites. These materials combine the low cost of natural fibers with the high performance of synthetic fibers, resulting in properties spanning a wide range. In addition, these materials show considerable promise in the PSA field.

Acknowledgments

This work was supported by the National Science Foundation, the U.S. Department of Energy, the United Soybean Board, the Maryland Soybean Board, the Delaware Transportation Institute, DuPont, John Deere & Company, Hercules Inc., and Cara Plastics Inc.

References

1. McCrum, N. G.; Buckley, C. P.; Bucknall, C. B.; *Principles of Polymer Engineering*, Oxford University Press: New York, 1997.
2. Devi, L. U.; Bhagawan, S. S.; Thomas, S. *J. Appl. Polym. Sci.* **1997,** *64,* 1739.
3. Bledzki, A. K.; S. Reihmane, J. Gassan, *J. Appl. Polym. Sci.,* **1996,** *59,* 1329.
4. Saha, A. K.; Das, S.; Bhatta, D.; Mitra, B. C. *J. Appl. Polym. Sci.* **1999,** *71,* 1505.
5. Ghosh, P.; Ganguly, P. K. *Plast. Rub. Comp. Proc. Appl.* 1993, 20, 171.
6. Gassan, J.; Bledzki, A. K.; *Polym. Comp.* **1997,** *18,* 179.
7. Gowda, T. M. Naidu, A. C. B.; Rajput, C. *Comp. Part A: Appl. Sci. Manuf.* **1999,** *30,* 277.
8. Shalash, R. J. A.; Khayat, S. M.; Sarah, E. A. *J. Petrol. Research* **1989,** *8,* 215.
9. Mishra; S.; Naik, J. B. *J. Appl. Polym. Sci.* **1998,** *68,* 1417.
10. Hargitai, H.; Czvikovszky, T.; Gaal, J.; Racz, I. In *Proceedings of the First Conference on Mechanical Engineering*, Budapest, 1998.
11. Hornsby, P. R.; Hinrichsen, E.; Tarverdi, K. *J. Mater. Sci.* **1997,** *32,* 443.
12. Mieck, K. P.; Luetzkendorf, R.; Reussmann, T. *Polym. Comp.* **1996,** *17,* 873.
13. Chen, X.; Guo, Q.; Mi, Y. *J. Appl. Polym. Sci*, **1998,** *69,* 1891.
14. George, J.; Sreekala, M. S.; Thomas, S.; Bhagawan, S. S.; Neelakantan, N. R. *J. Reinf. Plast. Comp.* **1998,** *17,* 651.
15. Rozman, H. D.; Kon, B. K.; Abusamah, A.; Kumar, R. N.; Ishak, Z. A. M. *J. Appl. Polym. Sci.* **1998,** *69,* 1993.
16. Kishore, R. M.; Shridhar, M. K.; R. M. V. G. K. Rao, *J. Mater. Sci. Lett.* **1983,** *2,* 99.
17. Shah, A. N.; Lakkad, S. C. *Fiber Sci. Tech.* **1981,** *15,* 41.
18. Williams, G. I.; Wool, R. P. *J. Appl. Comp. Materials.* **2000,** *7,* 421.
19. Morye, S. S.; Wool, R. P. *Polym. Comp.* **2000,** in press.
20. Wool, R. P.; Kusefoglu, S. H.; Palmese, G. R.; Zhao; R; Khot, S. N.; U.S. Patent 6,121,398, 2000.
21. Williams, G. I. M.S. Thesis, University of Delaware, 1999.
22. R. P. Wool and S. N. Khot, Proceedings ACUN-2, Sydney, 2000.
23. Wool, R. P. *Nature* **2001,** *409,* 772.
24. White, S. R.; Sottos, N. R.; Geubelle, P. H.; Moore, J. S.; Kessler, M. R.; Sriram, S. R.; Brown, E. N.; Viswanathan, S *Nature* **2001,** *409,* 792.

25. McCarthy, J. L; Islam, A. In *Lignin: Historical, Biological, and Materials Perspectives*; Glasser, W. G., Northey, R. A., Shultz, T. P., Eds.; Oxford University Press: Washington, DC, 2000.
26. Lin, S. Y.; Zhong, X. J. *China Pulp Paper* **1990,** *9,* 45.
27. Sarkanen, K. V.; Ludwig, C. H. *Lignins: Occurrence, Formation, Structure, and Reactions*; Wiley-Interscience: New York, 1971.
28. Kharade, A. Y.; Kale, D. D. *J. Appl. Polym. Sci.* **1999,** *72,* 1321.
29. DaCunha,C.; Deffieux, A.; Fontanille, M. *J. Appl. Polym. Sci.* **1993,** *48,* 819.
30. Barry, A. O.; Peng, W.; Riedl, B. *Holzforschung,* **1993,** *47,* 247.
31. Nada, A. M. A.; El-Saied, H., Ibrahem, A. A.; Yousef, M. A. *J. of Appl. Polym. Sci.,* **33**, 1987, 2915.
32. Ball, F. J.; Moorer, Jr., H. H.; Dougherty, W. K.; Canada Patent, Dec. 25, 1962.
33. Feldman, D.; Luchian, C.; Banu, D.; Lacasse, M. *Cell. Chem. Technol.,* **1991,** *25,* 163.
34. Wang, H. X. M. S. Thesis, Virginia Polytechnic Institute and State University, 1987.
35. Meister, J. J. Modification of Lignin by Free Radical Graft Copolymerization. *Proceedings of the 5th Annual Green Chemistry and Engineering Conference*, National Academy of Sciences: Washington DC, June 26–28, 2001.
36. Phillips, R. B.; Brown, W.; Stannet, V. T. *J. Appl. Polym. Sci.* **1972,** *16,* 1.
37. Yamamoto, H.; Amaiki, M.; Saitoh, H.; Sano, Y. *Mater. Sci. Eng.* **2000,** *7,* 143.
38. LaScala, J. J.; Wool, R. P. *J. Am. Oil Chem. Soc.,* Submitted.
39. Thielemans, W.; Can, E.; Morye S. S.; Wool, R. P.; *J. Appl. Polym. Sci.* **2001,** in press.
40. Bunker, S. P.; Wool, R. P.; U.S. Patent Application 2001.
41. Wool, R. P.; *Chemtech,* **1999,** *29,* 44.

Chapter 15

Toward a Green Chemistry and Engineering Solution for the U.S. Energy Industry: Reducing Emissions and Converting Waste Streams into Value-Added Products

M. M. Mercedes Maroto-Valer, John M. Andrésen, and Yinzhi Zhang

The Pennsylvania State University, The Energy Institute and Department of Energy and Geo-Environmental Engineering, 405 Academic Activities Building, University Park, Pennsylvania, PA 16802–2303

The U.S. electric power utility industry faces environmental challenges due to the emissions of nitrogen oxides (NO_x) by coal combustion furnaces and the associated increase in byproduct waste streams. This research program focuses on the development of a green chemistry and engineering solution for the U.S. energy industry that can guarantee a long-term source of energy for the 21st century by simultaneously reducing emissions and converting byproduct waste streams into value-added products. This solution combines the installation of low-NO_x burners that efficiently reduce NO_x emissions coupled with strategies that manage the associated increase in byproduct streams, mainly fly ash and unburned carbon.

Introduction

The U.S. electric power industry relies heavily on the use of coal as the main energy source, and coal-fired units generate over 55% of the total electricity produced annually in the United States (*1*). Coal is the most abundant fossil fuel resource in the United States, and therefore it will play an increasing role as a long-term energy source in the 21st century. Furthermore, carbon sequestration technologies are presently being developed and will be coupled with coal combustion systems. These technologies could potentially turn coal into a more environmentally-friendly energy source, as described below. Nevertheless, for coal to continue being a longer-term energy source, it must first meet current environmental challenges, such as emissions of pollutants like NO_x and the dissipation of waste streams. The implementation of Title IV of the 1990 Clean Air Act Amendments regarding NO_x emissions is mainly being addressed in coal combustion furnaces by installing low-NO_x burners. Although these technologies efficiently decrease NO_x emissions by lowering the temperature of combustion, they also reduce the combustion efficiency, with a corresponding increase in the concentration of uncombusted coal in the fly ash, generally referred to as unburned carbon (*2, 3*). Accordingly, this research program focuses on developing a green chemistry and engineering solution for the U.S. energy industry that can guarantee a long-term source of energy for the 21st century by simultaneously reducing emissions and promoting the commercial use of energy-related byproduct waste streams.

This solution combines the installation of low-NO_x burners that efficiently reduce NO_x emissions coupled with strategies that manage the associated increase in byproduct streams, mainly fly ash and unburned carbon. These strategies advocate the recovery and use of energy byproducts by developing two novel routes for the commercial utilization of unburned carbon present in fly ash as a precursor for premium carbon materials: (1) the use of unburned carbon as a precursor for the production of activated carbon, and (2) the use of unburned carbon as a substitute for calcined petroleum coke in the production of carbon bodies, as described below. The implementation of this research program could have a vast benign environmental impact that is closely related to the Pollution Prevention Act of 1990, because this program prevents dissipative pollution at its source by simultaneously reducing NO_x emissions and by reducing byproduct waste streams disposed of as waste. Furthermore, two indirect benefits from the implementation of these research strategies are associated with the potential application of activated unburned carbons for mercury capture and the reduction of CO_2 emissions from the high-energy demanding manufacture of cement, as described below.

The U.S. Energy Industry

The U.S. electric power industry relies heavily on the use of coal as the primary energy source, as mentioned above. In 1998, more than 950 million tons, or 90% of the total U.S. coal production, was used in coal-fired units to generate over 55% of the total electricity produced in the United States (*1*). Furthermore, the present domestic coal reserves are estimated to last at least over 250 years, while the current proven domestic oil and natural gas reserves are only expected to last less than ten years. These projections can certainly have a more optimistic side, considering that new reserves are being found and that more efficient systems and new technologies are being developed. However, under the current scenario, coal is expected to be a long-term domestic energy source for the United States.

Carbon sequestration technologies are presently being developed and could help to make coal a more environmentally friendly energy source. For example, the capture of CO_2 emitted from stationary combustion sources into natural sinks like forests can potentially close the carbon cycle by producing biomass, peat, and low-rank coals that could be used in combustion for energy production within a time frame of 15 to 100 years. The so-called "energy crisis" endured in the beginning of 2001 by Californians, as well as in other states, is indeed an unfortunate example of failed attempts to move away from coal energy while not having a viable substitute. Nevertheless, for coal to continue being a long-term energy source, it must first meet current environmental challenges caused by emissions of pollutants such as NO_x from coal combustion furnaces and by increases in the generation of byproducts.

Energy-Related Emissions and Waste Streams

Under Title IV of the 1990 Clean Air Act Amendments, the U.S. Environmental Protection Agency (EPA) issued a two-phase strategy to reduce NO_x emissions from coal-fired utility boilers. Phase I took effect in January 1996 and promulgated that emissions levels from Group-1 boilers should be reduced by over 400,000 tons/yr between 1996 and 1999 (*4*). Group-1 is formed by dry-bottom wall-fired boilers and tangentially fired boilers. Their emission levels were required to be reduced to 0.50 and 0.45 lb NO_x/MM Btu, respectively. Phase II, which began in January 2000, will achieve an additional reduction of about 900,000 tons of NO_x annually. Phase II has lowered the emission levels for Group-1 boilers to 0.46 and 0.40 lb NO_x/MM Btu for wall-fired boilers and tangentially fired boilers, respectively, and has established limits for Group-2 boilers, which include cell, cyclone, wet-bottom, and vertically-fired boilers, to levels of 0.68, 0.86, 0.84, and 0.80 lb NO_x/MM Btu, respectively (*5*). Nevertheless, Phase II regulation was initially dormant during a litigation process that was originated by eight upwind states and a number of utilities, which challenged EPA's authority to issue the rule (*6*). In February 1998, a Federal appeals court upheld EPA's rule, and the court denied the utilities' petition in its entirety. The administration has reactivated the rule, and it is expected that facilities in 19 states could be affected by the implementation of this regulation.

One of the more extensively used approaches for meeting Title IV of the 1990 Clean Air Act Phase-I has been the retrofitting of several hundred existing boilers with low-NO_x burners. EPA has projected that around 85 to 90% of the boilers in both Group 1 and 2 could meet the required NO_x reduction levels by installing low-NO_x burners (*5*). Other technologies, such as Selective Catalytic Reduction and Selective Non-Catalytic Reduction, are suitable complimentary approaches to meet the required NO_x reduction levels. The installation of low-NO_x burners changes the flame-temperature profile as well as the flame chemistry, because in essence, a hot, oxygen-rich flame is replaced by a cooler and longer fuel-rich flame. While these modifications have proven effective in reducing NO_x emissions, they also result in lower combustion efficiency, leading to an increase in the concentration of unburned carbon. This makes the ash unsuitable for use by the cement industry, because the unburned carbon tends to adsorb the air-entrainment reagents that are added to the cement to prevent crack formation and propagation (*2*).

Consequently, the carbon-rich ash is either placed in holding ponds or is landfilled, thereby increasing the potential damage to the environment from leaching. For instance, in 1999 approximately 60 million tons of ash, which is the main coal combustion byproduct, were generated. However, only one-third of all the ash was used, with the largest specialized application being for cement, due to the pozzuolanic properties of the ash (*7*). The remaining ash, over 40 million tons, was placed in landfills. Although in 1993 EPA determined that coal combustion byproducts are not hazardous, specially designed landfills are required to prevent the possible leaching of trace elements from the fly ash, which can reach drinking water sources. The landfill option also results in rising costs for the utility companies, paired with the loss of revenue from unsold ash. It has been estimated that the disposal of the approximately 20 million tons of ash, which are currently being marketed, will result in an additional cost of over $450 million to the utilities industry, which will correspondingly cause a rise in consumer electricity bills. Furthermore, environmental pressures regarding landfills will only intensify in this century, and to guarantee a domestic long-term energy source for the United States, the conventional processes for coal use have to be redesigned to comply with environmental regulations. Therefore, the increasingly severe regulations on landfills and the limited access to new disposal sites, together with the subsequent rise in the cost of disposal, will force the U.S. energy industry to recycle a larger amount of waste from coal combustion processes.

Green Chemistry and Engineering Solution: Simultaneous Reduction of Emissions and Use of Waste Streams as High-Value Precursors

The U.S. energy industry requires a green chemistry and engineering solution that can guarantee a long-term source of energy by simultaneously reducing emissions and byproduct waste streams. This can be achieved by installing low-NO_x burners, which have been proven to reduce NO_x emissions efficiently, coupled with strategies to manage the associated increase of byproduct streams, as illustrated in Figure 1.

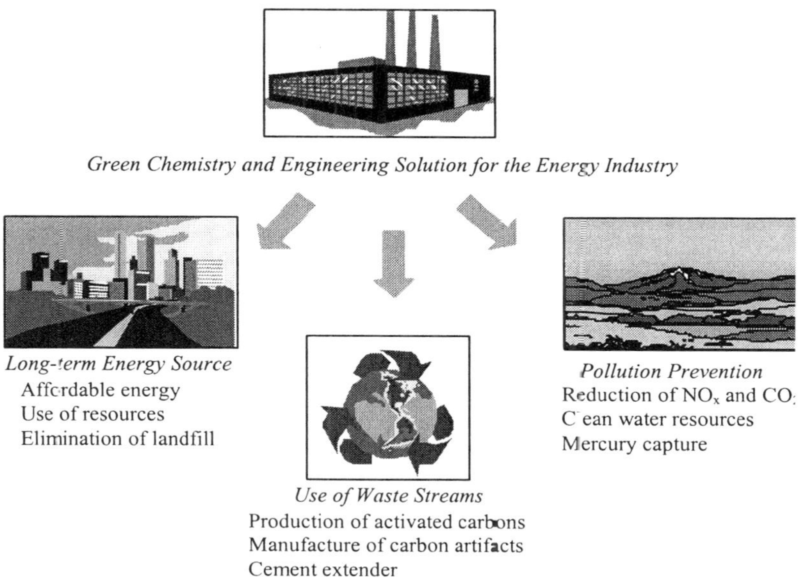

Figure 1. Representation of a green chemistry and engineering solution designed for the U.S. energy industry.

Accordingly, the recovery and recycling of energy byproducts can bring enormous economical and environmental benefits to the U.S. energy industry and to society at large. However, the implementation of Clean Air Act regulations is potentially detrimental to the marketability of ash, in the sense that more unburned carbon is present in the ash, with the subsequent rise in waste streams to follow. Several technologies have been successfully developed to recover and separate the unburned carbon from the ash. These technologies can be divided into wet (froth flotation) and dry (triboelectrostatic) processes, and they can produce inorganic ash concentrates with purities that meet the requirement for the cement industry, as well as carbon concentrates with purities of 85% and above (*2*). However, only a few power plants have installed a beneficiation process on their sites. One of the main reasons for this is the low value of the resultant separation products, since one ton of "clean" fly ash is generally sold for as little as $10, and the unburned carbon concentrated stream is simply rerouted to the combustor. However, the economics of this process can be significantly enhanced if both separated materials can be used as precursors for high-value products. Accordingly, this research program focuses on the development of novel routes for the commercial use of unburned carbon present in fly ash as a precursor for premium carbon materials: (1) activated carbons, and (2) carbon artifacts, as described below.

Unburned Carbon as Precursor for Activated Carbons

The present global consumption of activated carbons is over 350,000 tons, and it is estimated to rise 7% annually. The main reason for this expanding market is the ubiquitous use of activated carbons as adsorbent materials in a broad and increasing range of household, medical, industrial, military, and scientific applications (*8*). These include gas-phase adsorption in household air conditioning equipment and in industrial emissions control, liquid-phase adsorption for water treatment, and even gold recovery. Due to the expanding market for activated carbons, especially in applications related to environmental protection, new precursors are being sought. However, for these new precursors to compete effectively with conventional raw materials, such as wood, they must be inexpensive, have a low mineral matter content, and be easily converted into activated carbons. The unburned carbon in the fly ash satisfactorily furnishes all of these conditions, because it can be (1) easily obtained from the utility industries as a byproduct, (2) beneficiated from the fly ash by commercially available techniques; and (3) readily activated, because it has already gone through a devolatilization process while in the combustor and, therefore, only requires activation.

Currently, 6 million tons of calcined petroleum coke, worth around $1.2 billion, is sold annually in the United States for the manufacture of carbon artifacts. Although calcined petroleum coke is the dominant precursor for carbon artifacts, the unburned carbon from coal combustion is a potential competitor because of its thermal history (*9*). The unburned carbon has been treated at temperatures well above 1200 to 1300 °C and can be regarded as calcined coke. For example, the H/C atomic ratios of the unburned carbon are below 0.02, and they are comparable with commercial precursors for carbon artifact production, where the typical H/C atomic ratios are around 0.01. The challenging issue in this task will be to meet the purity requirement, because typical commercial petroleum cokes contain less than 1% ash. Sink/float methods using a high-density liquid media can effectively enrich the carbon content by reducing the ash to around 9 to 10%. Other methods, such as centrifugation, can further reduce this value. Therefore, the unburned carbon from coal combustion could be a strong competitor to petroleum coke for the production of carbon artifacts (*9*).

Experimental Procedures

Procurement of Fly Ash Samples

The fly ashes studied were collected from Shawville power station Unit 4 in Bradford Township, Pennsylvania, with a net capacity of 180 MW, and Portland station Unit 2 in Northampton County, Pennsylvania, with a net capacity of 243 MW, which are both operated by Reliant Energy. The hoppers were emptied prior to collection to obtain fresh ash. These units have been retrofitted with low-NO_x burners. Samples were collected all the way from the economizer through the hoppers to the stack, amounting to a total of 12 and 16 samples for Shawville Unit 4 and Portland Unit 2, respectively. The carbon contents, also called loss-on-ignition or LOI, of the fly ashes were determined according to the ASTM C 311 procedure. The fly ashes with the highest carbon concentration were chosen as feedstock for the production of the carbon products.

Activation of Unburned Carbon

The activation of the samples was carried out in an activation furnace, which consists of a stainless steel tube reactor inside a vertical tube furnace, as previously described (*8*). Typically about 2 to 3 g of sample was held isothermally at 850 °C for periods of 60 to 120 minutes in flowing steam. The porosity of the samples was characterized conducting N_2 adsorption isotherms at 77 K using a Quantachrome adsorption apparatus, Autosorb-1 Model ASIT. The Brunauer-Emmet-Teller (BET) surface areas were calculated using the adsorption points at relative pressures (P/P_0) of 0.05 to 0.25. The mesopore (pores 2 to 50 nm in width) and micropore (pores <2 nm in width) volumes were calculated using the α_s method.

Manufacture of Carbon Artifacts

The carbon pellets were produced from mixtures of unburned carbon, petroleum coke (either needle or sponge), and a coal tar binder pitch. The mixtures were heated to about 130 °C and pressed into pellets. The absolute densities of the precursors and the pellets were measured by using a Quantachrome MVP-1 Multi Pycnometer with helium as the density medium (*9*).

Results and Discussion

Carbon Variations between Hoppers

Fly ash samples were collected from each of the hoppers, and Figure 2 shows the configuration and gas flow for the hoppers of Shawville Unit 4 and the carbon values of the corresponding fly ashes, where the numbers on the bars indicate the corresponding hopper number.

The hot-side hoppers 13 through 16 presented the lowest carbon contents (12 to 17 wt %), while the cool-side hoppers 21 and 22 had the highest carbon contents (37 and 46 wt %, respectively). This “hot-side” and “cool-side” terminology has been adopted from former studies, and it is based on characteristics of the respective ashes, where fly ashes from hot-side collectors present characteristics associated with higher temperatures (lower carbon content and larger fly ash sizes) compared to those of their cool-side counterparts (*10*). The cool-side bins present the highest LOI values (40 to 50 wt %), and therefore they are suitable hoppers for the collection of high carbon content ashes as feedstocks for carbon materials precursors. Similarly, for Portland Unit 2, the hot-side bins (1 and 2) present LOI contents of only 10 to 12%, which become higher for hoppers 5 and 6 (15 and 19 wt %, respectively) and 9 and 10 (32 and 36 wt %, respectively) and reach a maximum for the cool-side hoppers 13 and 14 (36 and 50 wt %, respectively).

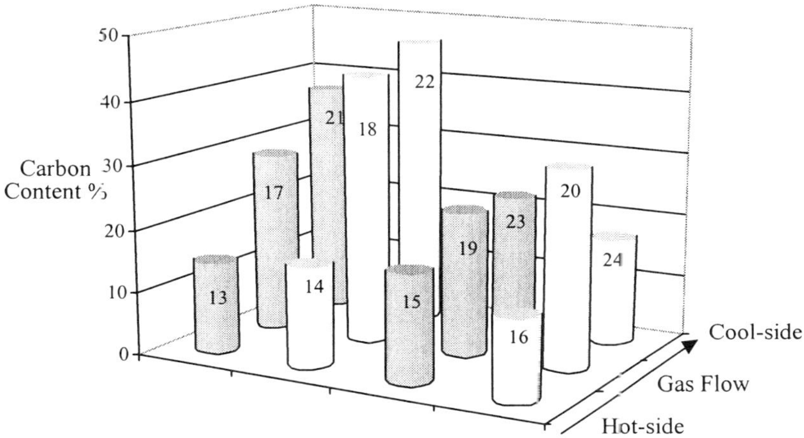

Figure 2. Variation in carbon contents of the fly ashes collected from the different hoppers of Shawville Unit 4.

Activation of Unburned Carbon Samples

Figure 3 shows the N_2-77 K adsorption isotherms for two of the unburned carbon samples investigated, designated as UC-A and UC-B, and illustrates the inherent porosity of these materials. Both adsorption isotherms are Type II according to the BDDT (Brunauer, Deming, Deming, and Teller) classification and are typical for nonporous or macroporous adsorbents, on which unrestricted monolayer-multilayer adsorption can occur (*11*).

Table I lists the BET surface areas (S.A.) for the precursors and their counterparts activated with steam for periods ranging from 60 to 120 minutes. The three UC parent samples have surface areas around 40 m^2/g. Previous studies conducted by the authors on a wide range of fly ashes and density gradient centrifugation concentrates have shown that the extensive and rapid devolatilization that coal undergoes in the combustor seems to promote the generation of meso- and macropores (*12*). Pore size distribution studies were also conducted and showed that the pore volume is mainly due to mesopores, with the mesopore volume accounting for over 60% of the total pore volume (*12*).

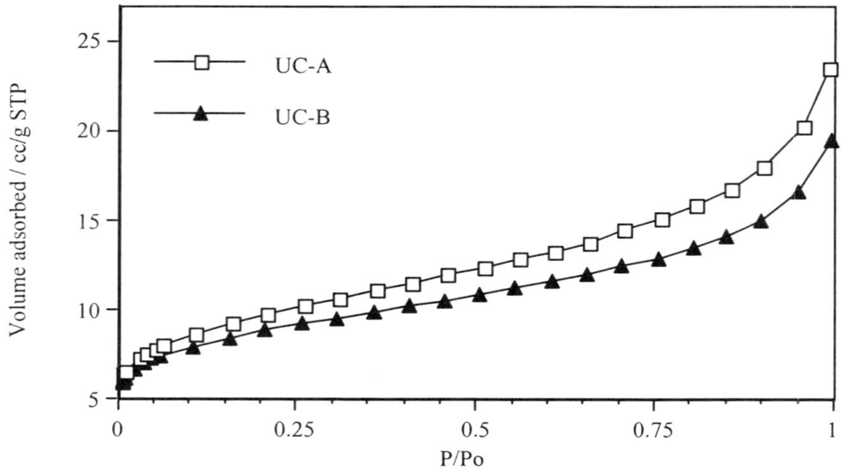

Figure 3. N_2-77 K adsorption isotherms of the two unburned carbon samples investigated.

Table I. Solid Yield and BET Surface Area for the Unburned Carbon Samples and Their Steam Activated Counterparts

Sample	*Activation Time (min)*	*Solid Yield (wt %)*	*BET S.A. (m^2/g)*
UC-A	—	—	40
UC-A	60	73	332
UC-A	120	35	540
UC-B	—	—	38
UC-B	60	55	443
UC-B	120	29	688
UC-C	—	—	40
UC-C	120	32	744

NOTE: The solid yields and surface areas are expressed in ash-free basis.

The solid yields of the UC samples activated for 60 and 120 minutes are also listed in Table I. Despite the low particle size of the UC samples, which typically is below 45 μm, the solid yields are relatively high, because the UC has already gone through a devolatilization process in the combustor (*8*). This makes the UC an attractive precursor for the production of activated carbons, because they present much higher solid yields than conventional precursors, such as wood. For the samples activated for 60 minutes, UC-A presents higher solid yields than UC-B (73% versus 55%), due to their larger particle size (200 µm versus 45 µm). Previous studies have shown that the particle size of the precursor strongly affects the solid yields of the resultant activated samples, with higher yields for bigger particle size fractions (*13*). As expected, the solid yields decrease as the activation time increase. The samples activated for the longest time (120 minutes) present higher BET surface areas than their counterparts activated for 60 minutes.

Figure 4 shows the N_2-77 K adsorption isotherms for the 60-minute steam activated UC samples. The isotherms are now more like Type I, with the typical concave shape to the P/P_0 axis, and they are typical for microporous materials. However, the isotherms do not reach a plateau at high relative pressures, indicating the presence of some meso- and macro-pores. Further activation studies are being conducted to ascertain the role of the ash in the activation process of the unburned carbon.

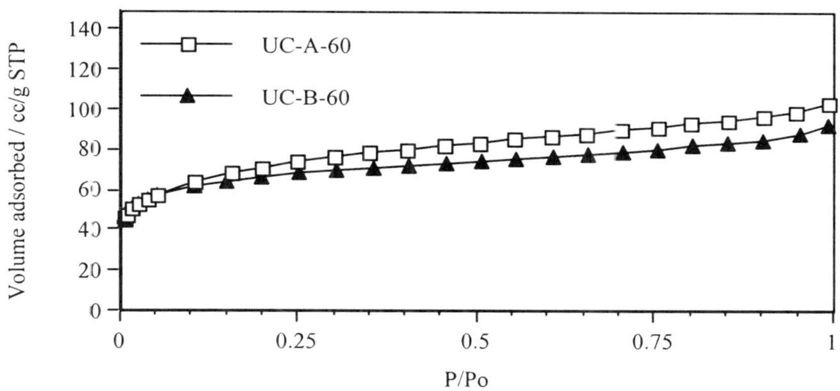

Figure 4. N_2-77 K adsorption isotherms for the activated samples.

Figure 5 compares the meso- and micropore volume for the precursors and the activated samples. As previously described, the inherent porosity of the UC samples is highly mesoporous, with the mesopore volume accounting for about 66% of the total pore volume. The activation process promotes the development of micopores, with the micropore volume now accounting for over 70% of the total pore volume. Samples were also activated using CO_2 and KOH, but the ones activated with steam presented generally higher surface areas than those using CO_2 activation or KOH pretreatment, probably due to the faster reaction rate of steam (*14*). Tests to assess the most suitable commercial applications of the activated carbons produced are currently being conducted in conjunction with industrial partners.

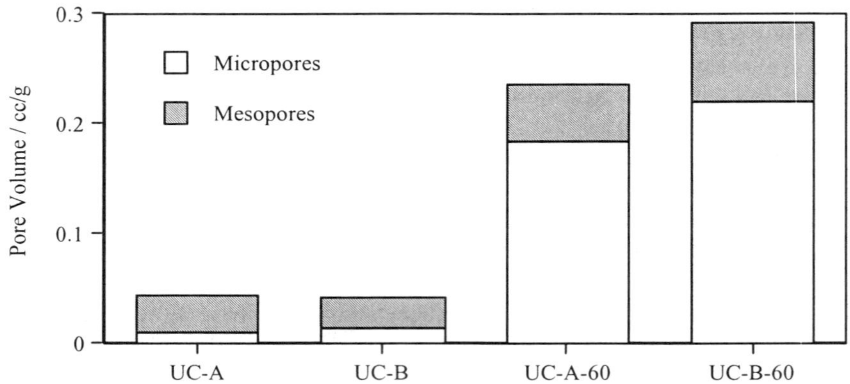

Figure 5. Distribution of the micro- and mesopore volume of the unburned carbon samples and their activated counterparts.

Properties of Carbon Pellets Prepared with Unburned Carbon

The conventional production of carbon bodies involves a filler, such as calcined petroleum coke; a binder, such as coal tar pitch; and a series of additives to enhance their properties. Carbon artifacts are used for a wide range of applications, including brushes for electrical machines, mechanical seals or brakes, artificial heart valves, anodes for aluminum smelting, and electrodes for steel arc furnaces, amongst others. In the preparation of conventional carbon materials, the petroleum coke is used as filler, and it is normally separated into at least three different particle size fractions: fine (<200 mesh, or 75 μm), intermediate, and coarse (>20 mesh, or 750 μm), to obtain improved packing densities. Due to the nature of the combustion process, most fly ashes are milled to at least the size of the fine fraction, which is used for the production of carbon materials. Hence, this study concentrated on the production of carbon pellets, where the fine fraction of petroleum coke was replaced by unburned carbon. Table II lists the formulations of the carbon bodies prepared with fly ash carbon, conventional calcined petroleum coke (either sponge or needle), and coal tar binder pitch prior to baking (i.e., green carbon bodies).

Table II. Formulations for the Carbon Bodies Investigated (wt %)

	Fine Coke	*Fine Unburned Carbon*	*Interm. Coke*	*Pitch*
CB-A	40 (Sp)	0	40 (Sp)	20
CB-B	20 (Sp)	20	40 (Sp)	20
CB-C	10 (Sp)	30	40 (Sp)	20
CB-D	40 (Ne)	0	40 (Ne)	20
CB-E	20 (Ne)	20	40 (Ne)	20
CB-F	10 (Ne)	30	40 (Ne)	20
CB-G	0	35	35 (UC)	30

NOTE: Sp = calcined sponge coke; Ne = calcined needle coke; UC = unburned carbon

Table III lists the baking yields and helium densities for the carbon pellets prior to (green) and after baking at 800 °C. The baking yields are very similar for all the carbon bodies investigated, around 90%, as is expected from the similar thermal history of petroleum coke and fly ash carbons (*3, 15*). The lower yield for the carbon body prepared using only fly ash carbon (CB-G) is lower than for the other bodies, 86.8% versus 90%, because of its higher pitch content compared to that of the other bodies investigated (30% versus 20%).

Table III. Baking Yields and Helium Densities for the Green and Baked Carbon Bodies Investigated

Carbon Body	*Baking Yield (wt %)*	*Helium Density (g/cc)*	
		Green	*Baked*
CB-A	90.2	1.791	1.976
CB-B	90.6	1.532	1.895
CB-C	90.2	1.523	1.893
CB-D	90.4	1.653	1.986
CB-E	90.5	1.541	1.959
CB-F	91.0	1.511	1.952
CB-G	86.8	1.551	1.842

As expected, the densities of the baked carbon bodies are higher than their green counterparts (Table III). The densities of the green baked carbon bodies produced with only petroleum coke, either sponge (CB-A) or needle (CB-D), are higher than those of the carbon bodies where fly ash carbon was used. This may be related to the lower density of the unburned carbon compared to that of calcined petroleum coke (*15*). However, upon baking, the densities of all the pellets containing unburned carbon are approaching that of the pellets containing only petroleum coke. Present studies are being conducted at our laboratories and focus on the optimization of the formulations, including a broader range of particle size distributions as well as mixing procedures, with the goal of maximizing the densities of the green carbon bodies produced. Other properties such as coefficient of thermal expansion are also being studied to determine the commercial use of the carbon bodies produced.

Economic and Environmental Benefits of the Implementation of the Proposed Green Chemistry and Engineering Solution

A hypothetical case study is presented here to evaluate the economic impact of the implementation of the above green chemistry and engineering solution for the U.S. energy industry. Consider a power plant with a unit size of 500 MW, which after installation of low-NO_x burners generates around 100,000 tons of fly ash annually, containing approximately 12% unburned carbon. The price for the sale of fly ash is about $10/ton, and the cost for disposal is $6/ton. Because the unburned carbon is not presently being marketed, its price has been estimated by comparing it to a similar carbon material, like anthracite, which sells for $80/ton, because both materials have low content of volatiles and very condensed structures. The criteria for the use of ash in the cement industry require carbon contents below 6% (ASTM C 618). Therefore, the ash of this hypothetical power plant can only be disposed of, incurring costs of $600,000. However, if the power plant separates the unburned carbon from the ash and markets these two products, as proposed in these green chemistry and engineering strategies, it will then obtain a total revenue of $1.84 million from the sales of both the ash ($880,000) and the unburned carbon ($960,000). Therefore, the implementation of the outcome of this research program will imply a total benefit of $2.44 million for this brief case study. The cost for the installation of a beneficiation process has not been included in this case study, but even though this installation may cost between $1 and 2 million, the revenues generated from the sale of the fly ash and unburned carbon will finance the initial investment for a separation process within the first year.

The direct environmental benefits of the implementation of this green chemistry and engineering solution are closely related to the Pollution Prevention Act of 1990, because this program prevents pollution at its source by reducing NO_x emissions though the use of low-NO_x burners coupled with the subsequent use of waste streams. Furthermore, two indirect benefits are associated with this implementation. The first one is associated with the potential application of the activated unburned carbons for mercury capture, because the capability of these carbons to capture mercury has been shown previously (*16*). The second indirect benefit comes through the reduction of CO_2 emissions from the high-energy-demanding manufacture of cement. Cement is an energy intensive product that requires about 1,500 megacalories (both fuel and electrical) per metric ton, or in other words, it consumes about two percent of the total fuel and electrical power worldwide. Therefore, by using fly ash as a cement extender as previously described, the amount of cement required to be produced is reduced, and therefore, the energy consumption in cement manufacture is also reduced with the subsequent decrease of CO_2 emissions from fossil fuel consumption.

Conclusions

The present work has focused on the development of a green chemistry and engineering solution for the U.S. energy industry that can guarantee a long-term source of energy for the next century by simultaneously reducing emissions and byproduct waste streams. This solution combines the installation of low-NO_x burners, which efficiently reduce NO_x emissions, coupled with strategies to manage the associated increase in byproducts streams, mainly fly ash and unburned carbon. The detailed study of a series of fly ash hoppers has revealed that cool-side bins present the highest LOI values (50%), and therefore they could be suitable hoppers for the collection of high carbon content ashes as precursors for carbon materials. This work has demonstrated that unburned carbon from coal combustion waste can generate activated carbons by steam activation, where after 120 minutes activation time, the unburned carbon samples generated activated carbons with surface areas above 750 m^2/g. Despite the low particle size of the samples investigated, the solid yields are relatively high, because the unburned carbon has already gone through a devolatilization process in the combustor. For the carbon artifacts, the baking yields are very similar for all the carbon bodies investigated, about 90%, as expected from the similar thermal history of petroleum coke and fly ash carbons. The densities of the baked carbon bodies produced with only petroleum coke were only slightly higher than those of the carbon bodies prepared using unburned carbon, indicating that a discrete optimization of the formulations could successfully produce commercial carbon artifacts from unburned carbon. Implementation of this green chemistry and engineering solution for the U.S. energy industry will also provide economic and environmental benefits.

Acknowledgments

The authors wish to thank the Consortium for Premium Carbon Products from Coal at The Pennsylvania State University and Reliant Energies for financial support.

References

1. U.S. Department of Energy. Energy Information Administration. http://eia.doe.gov/indexnjava.html.
2. Maroto-Valer, M. M.; Taulbee, D. N.; Hower, J. C. Novel Separation of the Differing Forms of Unburned Carbon Present in Fly Ash Using Density Gradient Centrifugation. *Energy Fuels* **1999**, *13*, 947–953.
3. Maroto-Valer, M. M.; Taulbee, D. N.; Hower, J. C., Characterization of Differing Forms of Unburned Carbon Present in Fly Ash. *Fuel* **2001**, *80*, 795–800.
4. U.S. Department of Energy, *Reducing Emissions of Nitrogen Oxides via Low-NO_X Burner Technologies*. Topic Report No. 5, 1996.
5. U.S. Environmental Protection Agency, EPA's Clean Air Market Programs; Acid Rain Programs URL: http://www.epa.gov/airmarkets/arp/index.html, Accessed June 2001.
6. Hogue, C., *Chem. Eng. News* **2000,** *78* (11), 10.
7. American Coal Ash Association. ACAA's Coal Combustion Products (CCPs) Production and Use Survey. http://www.acaa-usa.org/CCP_Survey1245.html (accessed June 2001).
8. Maroto-Valer, M. M.; Taulbee, D. N.; Schobert, H. H. Assessing Condensed Aromatic Coal Materials as Precursors for High-Value Carbon Products. *Prepr. Pap. Am. Chem. Soc., Div. Fuel Chem.* **1999**, *44*, 101–105.
9. Andrésen, J. M.; Maroto-Valer, M. M.; Andrésen, C. A.; Schobert, H. H. Fly Ash Carbon as a Filler for Carbon Bodies, *Proc. First World Conf. on Carbon* **2000**, *1*, 361–362.
10. Hower, J. C.; Rathbone, R. F.; Robl, T. L.; Thomas, G. A.; Haeberlin, B. O.; Trimble, A. S. Case Study of the Conversion of Tangential- and Wall-Fired Units to Low-NO_x Combustion: Impact on Fly Ash Quality. *Waste Management* **1997**, *17*, 219–229.
11. *Adsorption, Surface Area and Porosity;* Gregg, S. J., Sing, K. S. W., Eds.; Academic Press: New York, 1991.
12. Maroto-Valer, M. M.; Taulbee, D. N.; Hower, J. C. Characterization of Fly Ash Carbons Derived Due to the Implementation of NOx Clean Air Act

Amendments. *Prepr. Pap. Am. Chem. Soc., Div. Fuel Chem* **2000**, *45*, 401–405.

13. Maroto-Valer, M. M.; Andrésen, J. M.; Schobert, H. H. Determination of Structural Variations for a Series of Steam-Activated Anthracites. *Prepr. Pap. Am. Chem. Soc., Div. Fuel Chem* **2000,** *44*, 675–579.
14. Maroto-Valer, M. M.; Zhang, Y.; Andrésen, J. M.; Jones, A.; Morrison, J. L. Routes for the Development of Value Added Products from High Carbon Fly Ashes. *Prepr. Pap. Am. Chem. Soc., Div. Fuel Chem* **2001,** *46* (1), 316–317.
15. Maroto-Valer, M. M.; Andrésen, J. M.; Zhang, Y.; Andrésen, C. A., Morrison, J. L.; Schobert, H. H. Recovery and Utilization of Fly Ash Carbon, *Proc. Inter. Pittsburgh Coal Conference*, **2000**, CD paper 05-02.pdf.
16. Hower, J. C.; Maroto-Valer, M. M.; Taulbee, D. N.; Sakulpitkakphon, T. Mercury Capture by Distinct Fly Ash Carbon Forms. *Energy Fuels* **2000,** *14*, 224–226.

Indexes

Author Index

Subject Index

B

D

E

F

M

N

O

P

Q

R

T

U

Highlights from ACS Books

Desk Reference of Functional Polymers: Syntheses and Applications
Reza Arshady, Editor
832 pages, clothbound, ISBN 0–8412–3469–8

Chemical Engineering for Chemists
Richard G. Griskey
352 pages, clothbound, ISBN 0–8412–2215–0

Controlled Drug Delivery: Challenges and Strategies
Kinam Park, Editor
720 pages, clothbound, ISBN 0–8412–3470–

A Practical Guide to Combinatorial Chemistry
Anthony W. Czarnik and Sheila H. DeWitt
462 pages, clothbound, ISBN 0–8412–3485–X

Chiral Separations: Applications and Technology
Satinder Ahuja, Editor
368 pages, clothbound, ISBN 0–8412–3407–8

Molecular Diversity and Combinatorial Chemistry: Libraries and Drug Discovery
Irwin M. Chaiken and Kim D. Janda, Editors
336 pages, clothbound, ISBN 0–8412–3450–7

A Lifetime of Synergy with Theory and Experiment
Andrew Streitwieser, Jr.
320 pages, clothbound, ISBN 0–8412–1836–6

For further information contact:
Order Department
Oxford University Press
2001 Evans Road
Cary, NC 27513
Phone: 1-800-445-9714 or 919-677-0977
Fax: 919-677-1303